Drone

Drone

Remote Control Warfare

Hugh Gusterson

The MIT Press
Cambridge, Massachusetts
London, England

This book was set in Stone Sans and Stone Serif by Toppan Best-set Premedia Limited. Printed and bound in the United States of America.

Library of Congress Cataloging-in-Publication Data
Names: Gusterson, Hugh.
Title: Drone : remote control warfare / Gusterson, Hugh.
Description: London, England : The MIT Press, [2015] | Includes
  bibliographical references and index.
Identifiers: LCCN 2015039936 | ISBN 9780262034678 (hardcover : alk.
  paper)
Subjects: LCSH: Drone aircraft—History. | Drone aircraft—Moral and
  ethical aspects. | Drone aircraft—Social aspects. | Air warfare—United
  States—Moral and ethical aspects. | Air warfare—United States—
  History—21st century.
Classification: LCC UG1242.D7 G88 2015 | DDC 358.4/14—dc23 LC
  record available at http://lccn.loc.gov/2015039936

10 9 8 7 6 5 4 3 2 1

For my daughter Sasha
May you be a warrior for peace

The drone upsets the available categories, to the point of rendering them inapplicable.
—Gregoire Chamayou

Contents

Acknowledgments

This book was written when I spent a year at the Princeton Institute for Advanced Study. It is hard to imagine an environment that is more stimulating or more congenial to writing. Many colleagues at the Institute helped shaped my thinking, but six deserve special mention—Didier Fassin for his mentorship and remarkable breadth of knowledge and ideas; Joan Scott, who helped me think through the nature of "remote intimacy"; Michael Walzer, also writing on drones, whose questions forced me to think more deeply; Freeman Dyson, who, as smart as ever at ninety-one, is deeply committed to dialogue between the natural and social sciences and, for the founder of the company that makes the Predator and Reaper drones, is surprisingly skeptical of drone warfare; Richard Wilson, who in answer to a stray question over lunch about drones and the law, gave me an impromptu minilecture that provided the framework for my penultimate chapter; and Anver Emon, who was kind enough to review that chapter for legal accuracy and overall acuity.

Thanks also to Tom Boellstorff, Faye Ginsburg, and Michelle Murphy for productive suggestions about drones as a media technology and to old comrades from the Network of Concerned Anthropologists—particularly Catherine Besteman, Greg Feldman, and David Price—for being on the lookout for material for me. Katherine Chandler and John Cloud both helped improve my understanding of drone history.

This book was originally conceived as an elaboration of a chapter on the anthropology of drones that was written in response to an invitation from Matthew Evangelista and Henry Shue. My thanks to both of them for that invitation and their gentle insistence that I broaden my reading on the topic.

I tried out early versions of some of the arguments in this book in my regular column for the *Bulletin of Atomic Scientists*. A succession of editors there—Josh Schollmeyer, Mindy Kay Bricker, John Mecklin, Sasha Scoblic, and Elisabeth Eaves—gave me the latitude to experiment outside the normal territory of the *Bulletin* and helped improve my expression.

As this book simmered in my mind, I gave presentations to several audiences, whose members provided helpful feedback— the anthropology department at New York University, M. V. Ramana's series at Princeton's Global Security Program, Marie Chevrier's class at Rutgers University, a session at the International Studies Association organized by Stephen Schwartz, and refreshingly polyglot audiences at sessions organized by Greg Feldman at Simon Fraser University and Subrata Ghoshroy at the Left Forum.

Thanks also to Regula Noetzli for the kind gift of her negotiating skills, and Beth Clevenger for her editorial enthusiasm.

Finally, there is my family. Graham and Sasha were generous to let their father disappear to work on his book, and Jana Urbanova, our au pair, cheerfully stepped into the breach. My greatest gratitude goes to my wife, Allison Macfarlane. Not only did she read drafts of each chapter, but during the past year, I incurred such a childcare debt to her that we may need to have another child for me to repay it.

Institute for Advanced Study, Princeton, New Jersey
July 4, 2015

1 Drones 101

To the United States, a drone strike seems to have very little risk and very little pain. At the receiving end, it feels like war. Americans have got to understand that.

—General Stanley McChrystal[1]

Drone strikes generally take place on the edge of American public awareness. Although the United States has been using drones for what it calls "targeted killings" for over a decade, this was not formally acknowledged until a 2012 speech by Central Intelligence Agency (CIA) director John Brennan.[2] Even after that speech, CIA censors prevented Leon Panetta, the previous CIA director, from mentioning drone strikes in his memoir.[3] A U.S. State Department special envoy complained that "drones were a deeply classified topic in the government. You could not talk about them in public, much less discuss whom they were hitting and with what results. Embassy staffers took to calling drones 'Voldemorts,' after the villain in the *Harry Potter* series, Lord Voldemort: 'he who must not be named.'"[4]

Most people by now have a picture in their mind's eye of the drones themselves. The silver-gray planes have a signature bulbous nose and inverted V tail fins, while the planes' lack of windows lends them an eerie air of sealed-off blindness (figure 1.1). In the words of the *New Yorker*'s Steven Coll, they "look like giant robotic flying bugs."[5] Drone pilot Matt Martin calls them "extraterrestrial-looking."[6] Although the drones themselves have become an iconic image, media accounts of actual drone strikes are usually confined to a terse paragraph announcing that a strike took place, saying how many "militants" or al Qaeda members were killed, giving their names if they are known, and naming the faraway country where they died. An exception to this rule is a remarkably vivid account

Figure 1.1
A Predator unmanned aircraft.
Source: https://en.wikipedia.org/wiki/General_Atomics_MQ-1_Predator#/media/File:MQ-1_Predator_unmanned_aircraft.jpg.

published in the *New Yorker* of a drone strike in Pakistan as seen from the CIA command center:

On August 5th, officials at the Central Intelligence Agency, in Langley, Virginia, watched a live video feed relaying closeup footage of one of the most wanted terrorists in Pakistan. Baitullah Mehsud, the leader of the Taliban in Pakistan, could be seen reclining on the rooftop of his father-in-law's house, in Zanghara, a hamlet in South Waziristan. It was a hot summer night, and he was joined outside by his wife and his uncle, a medic; at one point, the remarkably crisp images showed that Mehsud, who suffered from diabetes and a kidney ailment, was receiving an intravenous drip.

The video was being captured by the infrared camera of a Predator drone, a remotely controlled, unmanned plane that had been hovering, undetected, two miles or so above the house. Pakistan's Interior Minister, A. Rehman Malik, told me recently that Mehsud was resting on his back. Malik, using his hands to make a picture frame, explained that the Predator's targeters could see Mehsud's entire body, not just the top of his head. "It was a perfect picture," Malik, who watched the videotape later, said. "We used to see James Bond movies where he talked into his shoe or his watch. We thought it was a fairy tale. But this was fact!" The image remained just as stable when the C.I.A. remotely launched two Hellfire missiles from the Predator. Authorities watched the fiery blast in real time. After the dust cloud dissipated, all that remained of Mehsud was a detached torso. Eleven others died: his wife, his father-in-law, his mother-in-law, a lieutenant, and seven bodyguards.[7]

There are a number of striking features to this account, which is told from the point of view of the executioners. A technology that is almost magical gives its owners, who are looking on the scene from high in the sky, a godlike power over life and death. The observation of the scene is simultaneously intimate and remote. It is also deeply asymmetrical: Mehsud, unaware of his exposure, is watched by faraway drone operators who can see him as if close up, reclining on the roof of his house on a hot evening as his wife attends to his medical needs. They get to frame the picture while he does not even realize he is in it. Without warning, he is killed as if by a god's thunderbolt from the sky. Seen from Virginia, the drone strike is quick, clean, and bloodless. Mehsud's death is instant. Nor, described unambiguously as a terrorist, does he seem undeserving of death. Twelve people die altogether, but the narrative marks only Mehsud's death as significant. The other deaths are almost outside the frame. And in a way that amplifies the strange mix of distance and intimacy, the scene is mediated entirely through a single sense—vision. The attack has no sound, smell, taste, or texture. And we are invited to experience it through a narrative of mastery and control—of the cool, righteous exercise of overwhelming power.

Contrast this with a very different account of a drone strike that was given by members of a Pakistani family who were invited (by a Congressman who opposes drone strikes) to testify to a sparsely attended congressional hearing. They described the shockingly sudden death of sixty-seven-year-old Momina Bibi. Testifying were her son, Rafiq ur Rehman, and her two grandchildren, nine-year-old Nabila and thirteen-year-old Zubair. This account is excerpted from a story in *The Guardian* newspaper:

"Nobody has ever told me why my mother was targeted that day," Rehman said. "Some media outlets reported that the attack was on a car, but there is no road alongside my mother's house. Others reported that the attack was on a house. But the missiles hit a nearby field, not a house. All of them reported that three, four, five militants were killed." He said, only one person was killed that day: "Not a militant but my mother."

"In Urdu we have a saying: 'aik lari main pro kay rakhna.' Literally translated, it means the string that holds the pearls together. That is what my mother was. She was the string that held our family together. Since her death, the string has been broken and life has not been the same. We feel alone and we feel lost."...

Rehman's son, Zubair, described the day of the attack, the day before the Muslim holy day of Eid, as a "magical time filled with joy." He told lawmakers that the drone had appeared out of a bright blue sky. ... "When the drone fired the first time, the whole ground shook and black smoke rose up. The air smelled poisonous. We ran, but several minutes later the drone fired again. People from the village came to our aid and took us to hospital. We spent the night in great agony in [sic] at the hospital and the next morning I was operated on. That is how we spent Eid." ...

His sister, Nabila, told lawmakers that she had been gathering okra with her brother and grandmother when she saw a drone and "I heard the dum dum noise. Everything was dark and I couldn't see anything. I heard a scream. I think it was my grandmother but I couldn't see her." ...

In testimony that caused the translator to stop and begin to weep, [Rehman] said: "As a teacher, my job is to educate. But how do I teach something like this? How do I explain what I myself do not understand? How can I in good faith reassure the children that the drone will not come back and kill them too?"[8]

This account is from the point of view of the victims, not the executioners. We share the experience of those who do not even realize that they are in the crosshairs until they are attacked. The account emphasizes the sudden incomprehensible eruption of violent force, literally out of the blue, in a warm scene of familial togetherness on an important holy day (figure 1.2). We are led to experience the drone strike through multiple senses, of which sight may be the least salient: we are told about the blackness of the smoke, the sound of the screaming, the smell of the explosion, the sensation of the ground trembling, and the pain of shrapnel wounds.[9] Unlike the first account, the narrative does not end shortly after the drone strike but dwells on the

Figure 1.2
The aftermath of a drone strike in Yemen.
Source: https://s.yimg.com/dh/ap/default/141113/drone_2.jpg.

aftermath—the physical pain of the survivors, the enduring grief
over the loss of the person "that held our family together."
Above all, this account foregrounds what is absent in the view
from CIA headquarters—the psychological suffering of those
on the ground, especially children, and the sense that the safe
predictability of life has been permanently destroyed. It is a narrative of helplessness, terror, and injustice. The drone operators'
perspective was remote and objectifying, but this narrative is so
affecting that it made the translator break down in tears.

In the decade and a half since the United States started using
weaponized drones, they have already begun to catalyze changes
in the nature of war. Drones are redefining what it means to be a
combatant, reshaping the sensory experience of war, and lever-
aging changes in operational tactics and military ethics. Above

all, drones pose urgent questions about the legal authority and political legitimacy of their use because they embody "an unconventional form of state violence that combines the disparate characteristics of warfare and policing without really corresponding to either."[10]

In this book, I explore the meaning and significance of this new kind of war, drone warfare, from multiple angles. Drawing on accounts by drone operators, drone victims, antidrone activists, human rights activists, international lawyers, journalists, military thinkers, and assorted academics, I ask how this new military technology is remaking the world. Although the public debate about drones has become increasingly polarized, my intent here is not to make a case for or against drones but to reframe the debate in ways that might freshen up the discussion. It is not my intent to give a comprehensive account of drone warfare, which would require a book three times as long. Instead, chapters 2 through 5 of this book present four substantive essays that tackle different aspects of drone warfare. Chapter 2, "War Remixed," looks at the way that drone warfare has redefined the space of the battlefield and considers the experiences of those who operate drones: What does their job involve? How do they subjectively experience the attacks in which they take part? What is the effect of war as commuter shiftwork on family life? Can drone operators be considered combatants? Chapter 3, "Remote Intimacy," examines the paradoxical mix of closeness and distance in the relationship between drone operators and their targets that can evolve over days of remote surveillance, and looks at what it is like to kill someone from over seven thousand miles away while watching as if close up on screen, whether it is easier or harder to kill someone this way than on the physical battlefield, and why drone operators seem to have high rates of posttraumatic stress disorder. Chapter 4, "Casualties," considers the debate between the U.S. government and its critics about the ethics and efficacy of drone strikes. Even as the U.S. government argues that drone technology offers the best hope for killing enemy combatants while sparing civilians, many human rights activists have claimed that drones have killed large numbers of civilians. Although I suggest that there has been a process of "ethical slippage" in the American use of drones, and that drones are killing more civilians than the U.S. government acknowledges, I make the case that the terms of the debate are misleading and that arguments of those for and those against drone strikes fail to recognize the ways in which the very distinction between civilians and combatants is under threat of collapse. Chapter 5, "Arsenal of Democracy?," probes the debate about drones and the international laws of war. It contrasts the arguments made by officials in the administration of President Barack Obama in favor of drone strikes with critiques from international lawyers and nongovernmental organizations, and it argues that drone warfare is a new form of state violence, hybridizing war and police action, that cannot easily be regulated by the international laws of war or by the checks and balances in the U.S. Constitution. It is reminiscent of old colonial practices and yet is different from anything we have seen before. As such, it invites debate over appropriate norms of action and control—a debate to which I hope this book will contribute.

A Brief History of the Drone

The online *Oxford English Dictionary* gives four principal definitions for the word *drone*—(1) noun: "the male of the honey bee ... a non-worker"; (2) noun: "a continued deep monotonous sound of humming or buzzing"; (3) verb: "to give forth a continued monotonous sound"; (4) verb: "to act or behave like a drone bee."[11] The first drone aircraft—radio-controlled biplanes intended as bombers—were developed during World War I but were unreliable and crashed often. In the mid-1930s the Royal Air Force, seeking to use a plane for target practice without killing pilots, developed a radio-controlled version of its Tiger Moth biplane. It was called the Queen Bee, and it seems that the word "drone" grew out of this name, partly through an analogy to lowly male drone bees that lacked stingers and eventually were killed by other bees.[12] Some drones were even painted with beelike black stripes on the tails.

In World War II, imitating the British example, the United States developed a new generation of drones for target practice in training exercises. The navy developed full-size drones while the army used smaller model airplanes. The United States also experimented with bombers from which the pilots would parachute to safety after takeoff, handing over control to pilots in other nearby planes. These "kamikaze" drones were packed with explosives so they could be crashed into military targets, both in Europe and in the Pacific. (During World War II, President John F. Kennedy's older brother, Joseph, died when one of these drones exploded prematurely, as so many did, before he could parachute to safety.)[13] Shortly after World War II, the U.S. military sent remote-controlled B-17s to collect samples from mushroom clouds in the first nuclear tests after Hiroshima and Nagasaki.[14]

In the 1960s, the U.S. military developed a new drone for surveillance purposes. Unlike today's drones, the Lightning Bug was jet-powered and preprogrammed. Usually launched from under the wing of a larger plane, this small drone flew a pro-grammed route, deployed a parachute as it fell, and was snatched from the air or collected at sea. Although it could fire missiles from the air, it did so only in tests. The plane was used for sur-veillance where manned flights would have been too dangerous, mainly over North Vietnam.[15]

During the 1973 Yom Kippur War, Israel demonstrated a novel use for drones. Israel was losing planes to the Egyptian air force, so it began to send aloft drones to draw Egyptian fire. Hav-ing thus used sacrificial drones to establish the location of Egypt's missile batteries, it sent manned planes to destroy them. Israel, which has historically been at the forefront of drone development, used drones again during its 1982 war in Lebanon to scout for targets.[16]

The United States purchased an Israeli drone, the Pioneer, to gather aerial intelligence during the 1991 to 1992 war in Iraq. Meanwhile one of Israel's leading drone designers, Abraham Karem, nicknamed the "Moses of modern drones" by one Penta-gon official, had moved to the United States, where he designed a new drone, the Predator.[17] Thanks to the development of global positioning system (GPS) technology and huge increases in the amount of data that could be relayed to and from drones via satellite, the Predator was the first drone that could be controlled from thousands of miles away. On the other hand, it was prone to crash in even moderately bad weather. First deployed in 1995, the new Predator drone was used by the United States to do reconnaissance during the Bosnian War and in 1998 to 1999 during the war in Kosovo. Its advantages over satellites for sur-veillance were immediately apparent. It could fly under the clouds that rendered satellites useless on many days, and unlike a satellite, which could take pictures only at those moments when it was orbiting overhead, the Predator could circle over-head for hours. "It's like having your own personal satellite over your target," said one military analyst.[18] The Predator did not yet carry missiles to attack targets on the ground, but at the end of the Kosovo War it acquired the ability to use lasers to "light up" ground targets for attack by manned planes.

The first combat use of a drone was by Iran in the Iran-Iraq War of the 1980s. Iran developed primitive drones that were used to fire rocket-propelled grenades at Iraqi ground forces.[19] Not until February 2001, however, when the United States test-fired a Hellfire missile from a Predator drone against a tank on the ground in Nevada, did it prove possible to launch a more substantial weapon from a drone without damaging the light-weight plane. (The Hellfire missile was chosen because it was light enough not to tear off the Predator's flimsy wings when fired).[20] In the words of historian Brian Glyn Williams, this "was a revolutionary moment in the history of aerial warfare. The unmanned reconnaissance drone had become a killer."[21] For different reasons, however, both the CIA and the U.S. Air Force were unenthusiastic about using it as such. Their reluc-tance to commit to what has become standard military practice just a decade and a half later speaks eloquently to the speed with which norms about remote killing have changed in only a few years. The CIA was banned from assassinating foreign leaders by a 1976 executive order that was issued by President Gerald Ford and had been observed by his successors in the Oval Office, and as recently as early 2001, CIA director George Tenet had qualms about the ethics and legality of killing people with armed drones. According to *The 9/11 Commission Report*, Tenet was "appalled" at the idea of assassinating people from the sky.[22] (Ironically, drones' ability to use lethal force with exact precision, thus sparing innocent civilians, put it on the wrong side of the assassination ban won by antiwar liberals in 1976.) Meanwhile, the air force, run by people who had made their careers in the cockpits of fighter jets and bombers, looked down on the lumbering, unmanned drones as unworthy of a first-rate air force. Furthermore, the U.S. government was on record as opposing "targeted killings" as practiced by Israel. As recently as 2001, Martin Indyk, U.S. ambassador to Israel, said that "the United States government is very clearly on record as against targeted assassinations. ... They are extrajudicial killings and we do not support that."[23]

Only a few months later, after the September 11, 2001, attacks, the U.S. position changed dramatically. In the words of Cofer Black, former director of the CIA's Counterterrorism Center, "There was 'before' 9/11 and there is 'after' 9/11. After 9/11 the gloves come off."[24] Ignoring the alarm to evacuate the CIA building after one of the hijacked planes crashed into the Pentagon, the CIA's chief lawyer spent the day of 9/11 with a yellow legal pad writing a first draft of a Memorandum of Notification, signed within a week by President George W. Bush, that authorized "targeted killings" of al Qaeda operatives and their allies.[25] Thus the United States began to use drones, along with special forces, to target selected individuals on the ground after it invaded Afghanistan in the fall of 2001. The Bush Administration hoped to kill Osama bin Laden, the leader of al Qaeda, or Mullah Omar, the leader of the Taliban, in this way.

The first drone strike was an inauspicious debut for the Predator. It was October 7, 2001, and the United States was using an armed Predator operated from a mobile command center in a parking lot in Virginia to track Mullah Omar, whose hideout had been identified. The CIA and the U.S. Air Force were running the mission together and, in the words of journalist Chris Woods, "the line of command which governed the Predator now shadowing Mullah Omar was blurred and untested." Air Force General Chuck Wald, who believed he was in charge of that night's operation, was in a command center in Saudi Arabia. He had two jet fighters holding twenty miles away, ready to strike Mullah Omar with 1,000-pound bombs on Wald's command. "The first I knew that Predator was there," said General Wald later, "was when I heard an unknown voice on my radio say 'You are cleared to fire.'" His deputy, General David Deptula, said he and Wald "both watched the weapon impact and both turned to each other simultaneously and said, "Who the fuck did that?'" Journalist Chris Woods reports that

Instead of striking the building in which Omar was thought to be located, a convoy vehicle was targeted and destroyed by the Predator and a number of bodyguards killed. ... When Predator pilot Scott Swanson finally broke his silence 13 years later to describe his role in the attack, he described the sight of a body thrown through the air as still "burned into my memory." The strike was aimed at drawing Omar outside: but in the chaotic moments that followed, the Taliban leader escaped.[26]

An attempt not long afterwards to kill Osama bin Laden with a Predator drone was no more successful. The CIA mistakenly believed its drone was tracking Osama bin Laden and two confederates because one of the figures on the ground seemed, like bin Laden, unusually tall. In the aftermath of the attack it became clear that the CIA had killed three humble peasants scavenging for scrap metal. The tallest, Daraz Khan, was five inches shorter than bin Laden. Pentagon spokesperson Victoria Clarke memorably explained that "we're convinced that it was an appropriate target ... [although] we do not yet know exactly who it was."[27]

Despite these inauspicious beginnings, the drone strikes continued. According to the Bureau of Investigative Journalism, Afghanistan has been targeted over a thousand times and "is the most heavily drone-bombed country in the world." The bombing was particularly heavy in 2012, when U.S. drone strikes in Afghanistan doubled in a single year as the United States drew down the number of troops on the ground. Most drones over Afghanistan have been operated by the U.S. military, rather than the CIA, and they have been deployed in concert with manned aircraft and ground troops to attack suspected Taliban forces. Roughly a quarter of all air strikes in Afghanistan have been drone strikes, and about 80 percent of drone strikes have been over Afghanistan and Iraq where drones would patrol above U.S. convoys, on the alert for ambushes or tell-tale signs of improvised explosive devices (IEDs) below.[28] I call this way of deploying drones, in which they are used to complement troops and other kinds of air power in an internationally recognized war zone, *mixed drone warfare.*

Drones were used in a similar way, as part of a mixed array of military force, in the initial U.S. invasion of Iraq in 2003 and against the evolving insurgency in Iraq in later years. U.S. drone use in Iraq declined precipitously after 2008, when President Obama, who opposed the initial invasion of that country, took office and began to withdraw U.S. troops. In 2008, there were sixty drones strikes in Iraq; in 2009, four; and in 2010, none.[29] After the abrupt rise of the new insurgency of the Islamic State in Iraq and Syria (ISIS), however, the United States recommenced drone strikes against Sunni insurgents in Iraq (then in Syria as well) in 2014.[30]

A different kind of U.S. drone campaign emerged in parallel in countries such as Yemen, Somalia, and Pakistan. The United States did not deploy large numbers of ground troops in these countries, and was not formally at war with them, but saw Muslim militant operations in them that it wanted to destroy. In some cases, particularly in Yemen, it sometimes had evidence that the militants were planning terrorist attacks on the United States, but in other cases, particularly in the so-called tribal areas of Pakistan, this was not the case. In these countries, drone strikes had a stand-alone quality. They were not part of a large war effort undertaken with F-16s, attack helicopters, and lots of boots on the ground, and they tended to be under the control of the CIA, or of the Joint Special Operations Command, rather than the air force. Because of this parallel drone campaign, the CIA underwent what the *Washington* Post called an "evolution from spy service to paramilitary force."[31] In the absence of a significant ground presence, the United States had the ability to attack unseen, from the blue, without presenting a reciprocal vulnerability to those it killed. By April 2014, over 6,800 U.S. troops had died in Iraq and Afghanistan, but none had been killed in Yemen and Pakistan.[32] Drone strikes in Yemen and Pakistan were the means of a truly asymmetric war in a way that was quite different from the situation in Iraq and Afghanistan. I call this stand-alone way of using drones without troops and other types of aircraft *pure drone warfare*. In pure drone warfare drones operate outside the legal framework that applies to an internationally recognized warzone, and the attention of the drone operators is focused primarily on the suspected insurgents they track and kill. This is different from mixed drone warfare in Afghanistan and Iraq, where drone operators track their own troops as well as insurgents and often report strong emotional involvement with their own troops on the ground who they are trying to protect from enemy fire. They may be as traumatized by their failure to save the lives of their own troops as by their own acts of killing. Although the media often portray pure drone warfare as the paradigm use of drones by the United States, most drone strikes have taken place in the context of mixed drone warfare in Afghanistan and Iraq.

The first drone strike in Yemen was in November 2002, and it killed Sinan al-Harithi, an al Qaeda leader thought to have been involved in the attack on the *USS Cole* in 2000. Five others died in the attack, including a U.S. citizen the CIA knew to be in the car with al-Harithi.[33] One target of the strike survived and was put on trial before a Yemeni military court, which acquitted him of terrorism charges, thus throwing into question the fastidiousness of U.S. targeting protocols. "The decision to kill al-Harithi and the vehicles' other occupants was a momentous one," says commentator Chris Woods. "After all, this was an assassination or targeted killing beyond the hot battlefield, exactly the kind of operation Bush officials had been so vociferously opposing with Israel. ... In a tetchy exchange with journalists two days later, [State department] spokesman Richard Boucher was still insisting that 'Our policy on targeted killings in the Israeli-Palestinian context has not changed.'"[34] Although it was not widely reported in the U.S. media, this targeted killing of someone in a country beyond the battlefield was condemned by some in the international community at the time. Sweden's foreign minister Anna Lindh described it "a summary execution that violates human rights," and added, "Even terrorists must be treated according to international law. Otherwise, any country can start executing those whom they consider terrorists."[35] The United Nations Human

Rights Council's special rapporteur, Asma Jahangir, called it "a clear case of extrajudicial killing" at odds with "international human rights and humanitarian law."[36]

After the 2002 strike, U.S. drones did not attack targets in Yemen again until 2011. This was partly because Yemen's president, Ali Abdullah Saleh, was furious when Deputy Defense Secretary Paul Wolfowitz embarrassed him by talking on CNN about Yemen's acquiescence in the strike.[37] But as Saleh's grip on power weakened and it became clear that al Qaeda was gathering strength in Yemen, the strikes resumed, now under President Obama. Another strike, on March 30, 2012, targeted Khadim Usamah, a doctor who was thought to be pioneering plastic explosives that were undetectable at airports and could be surgically implanted in suicide bombers.[38] By 2012 American drones were striking targets in Yemen about once a week.[39]

In 2011, the United States began a series of drone strikes in Somalia, which was being used by al Shabaab Muslim militants as a base of operations from which to conduct attacks in East Africa. At the time of writing, there have been about a dozen drone strikes in Somalia, where the United States has also deployed small special forces units and attacked targets with Cruise missiles.[40]

Additionally, there were 145 drone strikes in Libya during the upheaval there when Muammar Gadhafi fell from power in 2011.[41] And there were two drone strikes against insurgent training camps in the Philippines in 2006 and 2012 that were hardly reported in the United States.[42]

Pakistan has absorbed most of the drone strikes in this semi-covert drone campaign outside the public wars in Afghanistan and Iraq. In the words of the *New Yorker's* Steven Coll, "C.I.A.-operated drones waged what amounted to an undeclared, remotely controlled air war over North and South Waziristan, a sparse borderland populated almost entirely by ethnic Pashtuns."[43] This area, where annual per capita income averages $250 and only 17 percent of the population is literate, is one of the poorest and least developed in the world.[44] The United States was convinced, correctly, that Osama bin Laden, the person in the world they most wanted to find and kill, was hiding in Pakistan. As the United States began to lose ground in its counterinsurgency war against the Afghan Taliban, it became increasingly frustrated that Taliban fighters, mostly ethnic Pashtuns, crossed the border into the so-called tribal areas of Pakistan, sought refuge with the Pashtuns living there, regrouped, and returned to the fight in Afghanistan at a later date. Because the United States was not at war with Pakistan, it could not pursue these Taliban members across the border with ground troops or helicopters. Meanwhile Pakistan's government was increasingly worried about the Islamic insurgents who were gathering force in the semiautonomous area of Pakistan known as the Federally Administered Tribal Area (FATA), where they were enforcing a strict regime of Islamic law and plotting suicide attacks in other parts of Pakistan.

In a campaign led by the CIA, American drones began attacking targets in the "tribal areas" of Pakistan in June 2004. The pace of this bombing campaign picked up enormously in the final months of the administration of President George W. Bush and under President Obama, who doubled the rate of attack in 2010 after an al Qaeda suicide bomber blew up a CIA station chief and several other agents in Afghanistan. By 2011, the United States was carrying out one drone strike every three days on Pakistan.[45] To date, there have been at least 375 drone strikes on Pakistani territory, although their frequency fell dramatically around 2012.[46] The U.S. government said that its drones were killing only Islamic militants, but Pakistani and international human rights activists claimed that many civilians, including children, were being killed. Particularly incendiary was a 2006 strike on a school that Pakistani media reported as having killed seventy children. The strikes excited periodic protests in Pakistan, including one in Karachi in 2006 where 10,000 people chanted "Death to America!" and "Stop bombing innocent people!"[47] With opinion polls showing that as many as 90 percent of Pakistanis opposed the drone strikes and 74 percent of Pakistanis considered the United States to be an enemy,[48] Pakistan's government often denounced them. However, it was a public secret that was widely reported in both the U.S. and Pakistani media that Pakistan's leader had given permission for the attacks. Most of the strikes were conducted by drones based in Afghanistan, although the United States also operated for some years from a secret drone base in Pakistan whose existence Pakistan's government officially denied.

These attacks have been undertaken by two kinds of drones. The Predator was first deployed during the administration of President Bill Clinton, and the Reaper was first deployed in 2007 during the George W. Bush administration. "A little longer than the average station wagon," and weighing just 1,130 pounds, the Predator is surprisingly small. With its modified snowmobile engine it has a maximum speed of 135 mph, although it usually flies at speeds under 100 mph and, according to journalist Chris Woods, in Afghanistan on "some days the winds were so strong that the Predators would find themselves going backward."[49] The Predator can fly as high as 25,000 feet but usually is operated at 10,000 to 15,000 feet so that it gets better video imagery. It can stay aloft for about 24 hours at a time. About a quarter of the cost of the Predator goes into the "Ball," which is a rotating sensor ball that is mounted under the nose of the plane. It carries daylight cameras and infrared cameras that can pick up the outline of bodies at night, as well as equipment that scans the ground for cell phone signals, logging sim cards on the ground below. The cameras are said to be able to read a license plate from two miles up, and they feed data streams to controllers in different locations. According to one account, even filming from two miles up, the camera has "a lens so powerful it feels like a hawk hovering at 100 feet."[50]

The Predator is typically equipped with two Hellfire missiles for use against targets on the ground. Each missile costs around $70,000. The Predator flashes an infrared beam to "light up" or "sparkle" targets below that are then attacked by the Predator's missiles, by other planes, or by soldiers on the ground. These targets can be as small as individual insurgents who are fleeing an attack (what the U.S. military refers to as "squirters"), although the blast radius of a Hellfire missile is reportedly fifteen to twenty meters.[51]

The Reaper, a larger second-generation armed drone, can fly twice as fast as the Predator, go twice as high, and carry eight times as much ordnance. It is not limited to using the Hellfire missiles that force Predator operators to adopt a "one-size-fits-all" approach to bombing. Among other weapons, the Reaper carries the Small Smart Weapon, which is said to be capable of killing a person who is in one room while sparing people who are in the next room. It also has more sophisticated equipment for taking video footage of the ground and eavesdropping on electronic communications. Like the Predator, it is made by General Atomics.[52]

Both Predators and Reapers are launched from bases near the areas they patrol. They require substantial local crews to maintain them and local pilots who handle takeoffs and landings before passing control of the drones to crews in the United States. The American crews confer, often via typed chat, with a distributed network of people both in the United States and on the ground near the target zone. Although media coverage often makes it seem as if a drone needs only two or three people in a trailer in Nevada, the operation of a drone requires about 170 people in multiple locations. The people with their hands on the controls are the tip of a spear that extends from ground crews in Middle Eastern deserts to generals and lawyers in air-conditioned control rooms in the United States.[53]

The Appeal of Drones

Less than fifteen years after the first use of an armed drone by the United States, over 50 percent of the pilots being trained by the U.S. Air Force are drone pilots, and the proportion of remotely piloted aircraft in the U.S. fleet went from 5 percent in 2005 to 31 percent by 2012.[54] This is an extraordinary turnabout. Drones have proved attractive to the U.S. military for four principal reasons. First, they are far superior to both satellites and manned aircraft as tools for reconnaissance. Manned aircraft run out of fuel after a few hours, satellites pass over a site and then move on, but drones can linger over a location for a day or more, watching who enters and leaves a building or tracking the movements of people and vehicles that seem suspicious. They can also use infrared cameras to track people at night. And the video footage they generate can be archived so that it can be searched after attacks for signs of insurgent preparation. In such ways, drone surveillance helps in the mapping of insurgent networks and patterns of life as well as in locating arms caches and hiding places. The holy grail for drone advocates is a massive archive of drone surveillance footage that can be rewound so that analysts can work backward along an insurgent network—beginning with the explosion of a buried improvised explosive device (IED) and moving back to the insurgent who buried the device, the person from whom he collected it, and the bomb maker. So far, however, the enormous quantity, and often poor quality, of imagery has largely stymied attempts at this kind of data mining.[55]

Second, in the words of General David Deptula, "The real advantage of unmanned aerial systems is that they allow you to project power without projecting vulnerability."[56] Because the drone operator is safely ensconced in a trailer in Nevada, no American is killed or injured if a drone crashes or is shot down. This is beneficial in that the military does not like to see pilots killed, but also in the political sense that a war without American casualties is more likely to be a war without American opposition. Admiral Dennis Blair, former director of national intelligence, describes drone warfare as "politically advantageous." Saying that drone warfare enables a president to look tough without incurring American casualties, he adds, "It plays well domestically, and it is unpopular only in other countries."[57] In the words of British commentator Stephen Holmes, drones have "allowed the Pentagon to wage a war against which antiwar forces are apparently unable to rally even modest public support."[58]

Third, drones are cheaper than other aircraft, even after the costs of large support crews are considered, according to most analysts. Manned planes cost more to build because they have added features and redundant systems for the safety and comfort of their human occupants. (Drones, for example, have only one engine.) A Predator drone costs about $4.5 million, and a Reaper around $22 million. By comparison, an F-16 is about $47 million, and each new F-35 is projected to cost the American taxpayer between $148 million and $337 million.[59] And training a drone operator costs less than 10 percent of what it costs to train a fast-jet pilot.[60] Even though up to 50 percent of the U.S. Predator fleet has been involved in crashes, many of which destroyed the plane, they are still a bargain.[61]

Finally, their video surveillance capability and laser-guided munitions afford drones high levels of precision in the execution of attacks. Ground artillery certainly cannot match the precision of a Hellfire missile. Although other aircraft with laser-guided bombs may be able to achieve comparable levels of accuracy, the drone can linger for hours waiting for a good shot. Reportedly, this has been particularly important to President Obama. The *New York Times* said that "the drone's vaunted capability for pinpoint killing appealed to a president intrigued by a new technology and determined to try to keep the United States out of new quagmires. Aides said Mr. Obama liked the idea of picking off dangerous terrorists a few at a time, without endangering American lives or risking the yearslong bloodshed of conventional war."[62]

It is important to understand that the drone is not just a new machine that has been slotted into existing war plans in a space formerly occupied by other kinds of airpower. Instead, in concert with special forces on the ground, it is a pivot around which the United States has created a new approach to counterinsurgency warfare and border policing that is organized around new strategies of information gathering, precision targeting, and reconceptualizing enemy forces as a cluster of networks and nodal leaders. Advocates of this new paradigm argue that identifying and eliminating high- and midlevel leaders will cause insurgent networks to weaken and collapse. According to the journalist Andrew Cockburn, this new approach to counterinsurgency grew partly out of 1990s tactics in the war on drugs in Latin America, which focused on identifying and removing drug kingpins and cartel leaders. It was also strongly influenced by Israeli counterterror tactics that stressed the efficacy of arrests and targeted killings of first-, second-, and third-tier Palestinian leaders as part of a slow, relentless campaign to wear down Palestinian resistance. Referring to such a campaign as "mowing the grass," one Israeli intelligence official said, "All the time we have to mow the grass—all the time—and the leaders with experience will die and the others will be without experience and finally the 'barrel of terror' will be drained."[63] In U.S. counterinsurgency operations, this constant mowing of the grass is undertaken by drones that are operated by the CIA and the Pentagon and by elite ground forces from the Joint Special Operations Command (JSOC). JSOC increased in size from 2,000 people immediately before 9/11 to around 25,000 a decade later.[64]

Robert Scales, commandant of the U.S. Army War College from 1997 to 2000, described this new approach in a *Washington Post* column in which he contrasted the mechanized war approach that brought victory in World War II with what he calls "the McChrystal method," after General Stanley McChrystal—a former commander of JSOC:

Over the past 20 years, McChrystal and his teams have developed another uniquely American method of war by substituting skill, information and precision for mass, maneuver and weight of shell. We first watched the McChrystal method at work in Afghanistan following the attacks of Sept. 11, 2001, when small Special Forces units and the Afghan Northern Alliance teamed to destroy the Taliban using precision strikes delivered from aircraft high overhead. ...

In many ways, the McChrystal method is the opposite of shock and awe. It is often painfully deliberate, fed as it is by the patient collection of intelligence wrung from sources as disparate as informants and the big ears of the National Security Agency. Nothing happens without repetitive, realistic planning and rehearsals. No operation goes down without involving many layers of "enablers." Intelligence officers feed information constantly to teams as they move to the fight. Armed and unarmed drones feed video of enemy movements. Some of the killing is done up close, to be sure, but most comes from precision aerial weapons that obliterate the enemy in the dead of night.

Scales concludes by saying that "the Islamic State cannot be defeated by diplomacy, sanctions, coalitions or political maneuverings. Its fighters must eventually be killed in large numbers, and Americans will never allow large conventional military forces to take them on." That means we need more drones. Calling drones "the modern equivalent of Patton's tanks," he calls for the United States to amass ten times as many as it now has.[65]

Politicians, pundits, and military leaders portray the turn to drones as a sign of American strength. As one of the few countries with the technical sophistication and the infrastructure of satellites and military bases that are required to operate drones, the United States is now able to kill its enemies while remaining invulnerable. It is moving toward war that is so asymmetrical that only the other side will incur casualties, so asymmetrical that it is more like hunting than war. But another way of looking at this development is that American attempts to occupy Iraq and Afghanistan with ground forces or even to make a single U.S. assault force raid in Somalia in 1993 proved so disastrous in terms of military defeat on the ground and political opposition at home that the United States has been forced to retreat into the air and to cede the terrain it wants to control on the ground to the enemy. Drones have enabled improvements in aerial surveillance and in the interception of cell phone and radio signals on the ground, but insurgents have partly adapted to this by changing cell phones frequently, using couriers, spoofing aerial video cameras, and altering their meeting habits. Sometimes insurgents hide under bridges, where drones cannot see them, then change direction or switch cars. They also take advantage of urban topography, where cars may look alike or be hard to follow as they drive behind buildings, to elude surveillance.[66] On occasion, adversaries have also succeeded in hacking U.S. drones. In 2009, Shia insurgents in Iraq used software available for $29.95 on the Internet to hack into drone video feeds that were not encrypted so that they could use U.S. drone footage for their own battle planning. More seriously, in 2011, Iran succeeded in capturing a U.S. RQ-170 surveillance drone by hacking into its communications and reprogramming it to land—intact—within Iran, where it was promptly put on display to the international media.[67] Because they are powered by a single weak engine, drones are more vulnerable than manned planes are to strong winds, and they cannot fly in bad weather. Their slow, lumbering frames lack maneuverability, so they are easy to shoot down with ground-to-air or air-to-air missiles (as the Serbs demonstrated over Bosnia in the 1990s). "Pick the smallest, weakest country with the most minimal air force—[it] can deal with a Predator," says U.S. Air Force General Mike Hostage.[68] And it hardly matters that insurgents in Afghanistan and Iraq lack the ability to shoot down drones given that around 50 percent of the Predators have crashed by themselves.[69] Unlike ground forces, drones cannot search houses, interrogate captives, talk to villagers while patrolling their streets, win hearts and minds through development projects, build relationships with tribal elders, or take physical custody of territory. All they can do is track people and kill them. But effective counterinsurgency requires more than just killing insurgents, and a country that has been chased offshore into the air may not be capable of effective counterinsurgency, no matter how many PowerPoint slides they have about "vertical power."

As is shown in the following pages, although drones offer unprecedented accuracy in destroying targets on the ground, the targets sometimes turn out to have been misidentified, and when many civilians are killed along with the intended victim, the cost in local anger may outweigh the benefits of killing another midlevel insurgent who will soon be replaced. And the more people the drones kill, the longer the target lists seem to become.

Ever since General Giulio Douhet claimed in the early twentieth century that wars would now be won from the air, advocates of air power have repeatedly prophesied the imminent obsolescence of ground forces, but their prophecies remain as yet unfulfilled. In the words of the Israeli Eyal Weizman, "The fantasy of a cheap aerial occupation, or 'aerially enforced colonization,' is … as old as air forces themselves."[70] But as former U.S. Air Force pilot Shane Riza writes, "Sole aerial efforts at *controlling*—the word choice is important—populations or militaries on the ground have not worked ever since the British first tried it in Iraq in the 1920s."[71] Thus, in the pages that follow, as well as inquiring into the experience of those who fly drones and probing the implications of drones for democratic governance in the United States, we must ask the question almost all commentators conspire to bury: are these alleged new wonder weapons an effective tool at all for achieving the goals of the American national security state?

2 War Remixed

I have a team now
A headset full of backseat drivers
Analysts advisers JAGs
My lone wolf days are over I don't call the shots anymore
I'm like the backup singer for Neil Diamond.
—Drone pilot, in George Brant, *Grounded*[1]

By separating the pilot from the plane and shifting combat from an embodied to an onscreen experience, drone technology has remixed war. A new experience of war has been created by putting operators in desert trailers thousands of miles away from the planes they control and the people they kill, by allowing them to watch combat on screen, and by embedding decisions to kill in networks that span the world. In this chapter and the next, I map the nature of this experience, probing its implications for traditional concepts of delimited battlefields, honor and bravery in battle, and perceptions of the enemy. In doing so, I draw on many sources but particularly on the vivid memoir, *Predator*, written by the drone pilot, Matt Martin (with Charles W. Sasser).[2]

Although this chapter focuses primarily on the experience of the drone operators—the "stick monkeys" in windowless trailers in the Nevada desert and elsewhere who work twelve-hour shifts putting "warheads on foreheads"—it is important to recognize at the outset that drones are controlled by distributed networks of people. According to *Foreign Policy* magazine, it takes 168 people to keep a Predator in the air.[3] The person in the air conditioned trailer with a hand on the joystick is but one node, albeit a central node, in that network.

The largest node in the network (and the most invisible in media accounts) is the team of around seventy people who are responsible for maintaining the planes, getting them into the air, and landing them safely from airfields near their patrol routes. These teams—located on U.S. bases in Turkey, Sicily, Afghanistan, Djibouti, Ethiopia, Niger, Saudi Arabia, the United Arab Emirates (UAE), Uzbekistan, Chad, Cameroon, the Seychelles, and elsewhere[4]—include contractors, maintenance technicians, and pilots (figures 2.1 and 2.2). They do maintenance work and attach new missiles to the drones' wings before each flight. Pilots on these teams are responsible for takeoffs and landings, which are handled locally because the two-second delay in communications between an operator in the United States and a drone in the Middle East increases the risk of a crash during the delicate moments when the plane takes off or touches down. Even so, according to drone pilot Matt Martin, landing a drone "was more difficult than landing a conventional aircraft."[5] The fixed camera on the nose of the drone affords only a 30 degree view rather than the 50 degree view a pilot has from a cockpit; its long wings make a drone more susceptible than manned planes to turbulence and wind gusts; and the lack of

Figure 2.1
U.S. Air Force 46th Expeditionary Aerial Reconnaissance Squadron, July 2, 2004. Unloading a rocket from a Predator unmanned aerial vehicle at Balad Air Base, Iraq.
Source: Photo by U.S. Air Force Staff Sergeant Cohen A. Young, http://cryptome.org/2012-info/drone-crew/drone-crew.htm.

bodily feedback that is inherent in remote piloting is disconcerting for experienced pilots:

If you were actually inside a cockpit, you saw the altitude of your plane through direct and peripheral vision, heard the spool-up of the engine, felt the ground rush, and knew instinctively when to flare and ease the airplane onto the runway.

None of that applied to the Predator. You looked through the video camera "soda straw" with almost no peripheral view. You saw only what was square ahead, your view fixed to the nose. If you had to crab into a crosswind, your view crabbed with you. You couldn't hear the engine,

Figure 2.2
U.S. airmen with the 380th Expeditionary Aircraft Maintenance Squadron prepare an RQ-4A Global Hawk unmanned aerial vehicle aircraft for takeoff at an undisclosed location in Southwest Asia, December 2, 2010. *Source:* Photo by U.S. Air Force Staff Sergeant Eric Harris, http://cryptome.org/2012-info/drone-crew/drone-crew.htm.

couldn't feel the ground rush. Everything was pure instrument and video interpretation.[6]

As the drone moves out of the local team's line of sight, control is transferred to a team in the United States that operates the drone by sending commands through fiber optic cables across the U.S. and under the Atlantic to a facility in Europe that communicates with the drone via satellite. The drone uses the same route in reverse to send back video imagery of the ground below and readings of the drone's speed, altitude, fuel consumption, and so forth. Sometimes these communication links go down, and over 25 percent of drone crashes are connected to communication outages. But "usually, the outages last only a few seconds and are harmless. Just in case, drones are programmed to fly in a circular pattern until the links are restored. In worst-case scenarios, they are supposed to return automatically to their launch base."[7]

A drone is flown from a trailer called a ground control station (GCS), which one journalist likened to a "souped-up shipping container."[8] "Inside that trailer is Iraq, inside the other, Afghanistan," one visitor to a U.S. Air Force base was told.[9] According to another journalist, the air in the "hermetically sealed control center ... was suffused with traces of cigarette smoke and rank sweat that no amount of Febreze could mask."[10] This ground control station, eight feet wide and thirty feet long, is fitted with padded seats, several computer screens, and the controls for operating the drone. Inside is a team of three—a pilot, a sensor operator, and an intelligence analyst. (In various parts of the network, more intelligence analysts pore over the drone footage and electronic intercepts, and images are stored so that analysts can go back to them and hunt for the person who planted a bomb that later exploded, for example.)[11] The pilot uses a joystick to steer the drone and a lever to control its speed. Here is drone pilot Matt Martin's description of the scene in his trailer:

The GCS featured four screens in front of both the pilot and the sensor operator, and a ninth screen mounted between them. The tracker, the top screen in front of me, displayed a map of Baghdad and an icon to indicate the location of my plane. Below that the heads-up (HUD) screen showed video from the aircraft targeting pod. ... Below that, by the flight controls, were two heads-down displays (HDD 1 and 2), which showed engine and fuel data and other aircraft information. The mission computer, between me and my sensor operator, was split down the middle: the right half was a map showing the positions of all U.S. air and ground forces in the vicinity as represented by appropriate icons, the left half was divided into several "chatrooms" like those on the internet. Except these chatrooms functioned through a separate classified military internet.[12]

The journalist Matthew Power has described the scene in another ground control station, the center of operations for Brandon Bryant, who is the drone pilot on whom Andrew Niccol's Hollywood movie *Good Kill* is loosely based: "The pace of work in the box unraveled Bryant's sense of time. He worked twelve-hour shifts, often overnight, six days a week. ... A loaded Predator drone can stay aloft for eighteen hours, and the pilots and sensors were pushed to be as tireless as the technology they controlled. ... Mostly the drone crews' work was an endless loop of watching: scanning roads, circling compounds, tracking suspicious activity. ... Usually time passed in a haze of banal images of rooftops, walled courtyards, or traffic-snarled intersections. ... He mastered reading novels while still monitoring the seven screens of his station, glancing up every minute or two before returning to the page."[13]

It is a challenge to integrate data from these disparate screens. In the words of drone pilot Bill "Sweet" Tart: "It's an immense mental task to build a three-dimensional picture of your aircraft, over a target or operating with other aircraft, using a camera, using chat and text, using dials and gauges, using both an overhead look and a side view of the world, and integrating all that in your mind" (figure 2.3).[14] One commander of a drone squadron observed that younger pilots who grew up playing video games seem to adjust more easily to this split-screen multitasking: "They will sit there and watch all four

Figure 2.3
U.S. Air Force 46th Expeditionary Aerial Reconnaissance Squadron Predator pilots operate individual Predator unmanned aerial vehicles using remote controls at Balad Air Base, Iraq, July 2, 2004.
Source: Photo by U.S. Air Force Staff Sergeant Cohen A. Young, http://cryptome.org/2012-info/drone-crew/drone-crew.htm.

of their screens at once, monitoring everything from the map to the weapons to the fuel, while also peeking over at the pilot besides them's screen, to see what he is looking at. That comes from all those games. An older pilot like me was taught to go through the checklist one by one. He will look at each screen at a time."[15]

Drones often patrol over sites from which insurgents have attacked or been observed caching weapons in the past, and they relay video footage, stamped in the center with crosshairs, of buildings, vehicles, or individuals that have been "nominated" for killing by intelligence sources or that seemed suspicious from surveillance. The video footage seen by the crew controlling the drone is often piped simultaneously to military commanders at bases elsewhere, including the 609th Air and Space Operations Center in Qatar; teams of intelligence analysts in the United States; military officers and intelligence specialists in the vicinity of potential targets in, say, Afghanistan; and military air traffic controllers. The amount of data generated and archived by this surveillance system is staggering: "a thousand hours *per day* of full-motion video (defined as 24 frames per second) plus further streams of intercepted calls and associated signals intelligence that the current 5,000-strong complement of air force Distributed Common Ground System analysts will never have time to review."[16]

If someone on the "kill list" is being tracked or if U.S. forces are under attack on the ground beneath a drone, the chatrooms and radio wavelengths can spring to life with messages bouncing back and forth between nodes in the network as people try to make sense of the video footage from the drone and agree on a course of action. The crew that is operating the drone may suddenly find itself in the thick of exchanges—with air traffic controllers who are requesting that the drone keep its distance from manned planes moving in to attack, with junior officers on the ground who are urgently seeking help from the air against insurgents attacking them, with the pilots of manned planes asking the drone to "light up" a target for them with its lasers, and with senior commanders in the United States who are demanding to know whether that person on the ground is an insurgent with a mortar or a farmer with a shovel before they approve an attack.

Peter Singer, an analyst at the Brookings Institute, describes the confusion that this can generate:

Rather than relying on the judgment of their highly trained officers, generals increasingly want to inspect the situation for themselves. It's all fine if the enemy plays along and gives that general hours to watch the video himself and decide which bomb to use. But sometimes matters aren't decided on a general's schedule. ... The traditional concept of a military operation is a pyramid, with the strategic commander on top, the operational commanders next, and the tactical commanders on the bottom layer. With the new technologies, this structure isn't just being erased from above, with strategic and operational commanders now getting into the tactical commanders' business. It is also endangered from the sides. As one drone squadron officer explains, a major challenge in the command and control of reachback operations is their simultaneous location in multiple spaces. ... The results are "power struggles galore." As the operations are located around the world, it is not always clear whose orders take priority.[17]

Drone pilot Matt Martin gives voice to the frustration that this dispersed structure of decision making can create. One day, two members of his crew tracked "a boat chugging up the Tigris River":

They followed the boat to a weapons cache concealed in thickets along the river. The situation warranted a shot, but again I couldn't get a quick decision for them. The army brigade in charge of the area around the base wouldn't clear it until they saw the video and confirmed that the target was indeed hostile. That meant piping the video feed to the army's tactical operations center (TOC) so the battle captain could view it and decide on a course of action. By the time all that had occurred, the bad guys had stashed their load of weapons in the thicket and were clearing out of the vicinity. The bad guys got away because of stove-piping.[18]

As drone crews zero in on a human target (sometimes referred to as an "objective"), they are supposed to consider which missile is most likely to destroy the target while minimizing civilian casualties. Sometimes they weigh options with the aid of a computer program called *Bugsplat*, which calculates the likely destruction with various missiles and angles of attack. (A strike that kills many civilians is described as causing "heavy bugsplat.")[19] Military lawyers with expertise on the laws of war also participate in discussions about civilian casualties from strikes under consideration.

Here is Matt Martin's description of his first kill. He had been following an insurgent with a mortar, whom he nicknamed "rocket man," as he weaved through traffic in Sadr City, Iraq, in an old Ford. The insurgent eventually parked in the courtyard of a house. Martin and his joint terminal attack controller (JTAC), a soldier who is on the ground directing air attacks, worried that a drone strike would kill civilians nearby: "We had to be cautious with a shot in this neighborhood to avoid killing a bunch of people who didn't necessarily deserve being killed." At the same time, "the rocket man *deserved* being killed if anyone did." Even after he was "cleared hot" (authorized to strike) by the JTAC, Martin kept circling in the hope that "rocket man" would move back onto the open road before he called "rifle"— the order to shoot. When it became clear that he would run out of fuel if he waited much longer, Martin

began preparations for a shot by scrutinizing the target from all angles in order to choose the best approach to minimize collateral damage. I calculated that if I dropped one right down the middle of the yard on top of the Ford, the brick wall would buffer the explosion and leave adjacent houses relatively undamaged. ... I doubted whether B-17 and B-29 pilots and bombardiers agonized over dropping tons of bombs over Dresden or Berlin as much as I did over taking out one measly perp in a car.

I flew the Predator out to about twelve klicks [slang for *kilometer*], then turned inbound for the run. ... Senior airman Juan Abado, my sensor, armed his targeting laser. It was his job to guide the missile to its target once I fired it. If his hand twitched at the last instant, if he breathed wrong, the missile might go astray and take out the house full of people next door. ...

At 10 klicks out, I ordered Abado to begin lasing. At 8.9 klicks, I initiated my three-second countdown to "rifle" at 8.7. ... Concentrating on crosshairs superimposed on the target, I drew in a deep breath, felt sweat stinging in my eyes, tasted the bile of excitement in my mouth. I took a last look at the street in front of the house to make sure it was clear. Then I squeezed the trigger. The image on my screen pixilated as the airplane yawed from the asymmetric thrust of the missile's launching.[20]

Many drone operators describe the fifteen to thirty seconds it takes the missile to reach its target as an eternity: "Time became almost ductile, the seconds stretched and slowed in a strange electronic limbo."[21] They often report a sense that everything is in slow motion as they pray that a child does not step into the picture or that the target does not step out of range (insurgents reportedly exploit the four-second delay in communication to and from the United States by moving around a lot if they become aware that a drone is targeting them).[22] In Martin's case:

At the last moment, with only seconds to spare, the unthinkable occurred. An elderly man appeared on the screen, tottering along in front of the wall in his traditional Arab garb. ...

"Oh, Jesus! Move!"

My infrared screen flashed bright as the heat from the explosion washed across camera lenses. The image re-formed almost instantly to reveal that the missile had detonated precisely where I aimed it. Flames from the car were already crackling. ... As for the rocket man,

I doubted if even seventy-two virgins in Paradise would be able to put him back together again. ... The Rocket Man had it coming. The old man did not.[23]

When a drone attacks, "it is not this light precise pinprick that many Americans believe," says *New York Times* reporter David Rohde. Rohde had the misfortune to be held hostage by the Pakistani Taliban for over six months, during which time "the buzz of a distant propeller [was] a constant reminder of imminent death." A drone strike kills people on the ground by incinerating them, generating a blast wave that crushes internal organs, and unleashing a shower of high-velocity steel shrapnel. Speaking of a nearby drone strike that he survived, Rohde says, "it was so close that shrapnel and mud showered down on the courtyard. Just the force and size of the explosion amazed me. It comes with no warning and tremendous force."[24]

In the parlance of drone operators, the moment of missile impact is called "splash," which is an oddly liquid term for a highly kinetic event that shatters concrete, twists metal, and ignites fires. The drone continues to circle after the smoke clears so that the crew can evaluate what was destroyed, how many people died, and whether any of the dead appear to have been civilians. Counting and categorizing the dead is a challenge because they are often in pieces. At this point, drone crews also look for surviving insurgents who might have run off, and they target them with a second missile. Such runners are called "squirters" because it is assumed that they urinate on themselves in terror.[25]

For drone crews, the surveillance, pursuit, and destruction of their quarry are intensely visual experiences. The crew does not feel the recoil of the plane as the missile is released, does not hear the explosion or the screams that follow the strike, and does not inhale the fumes of burning fuel, flesh, and rubber. (George Brant's fictional drone pilot calls the climax "a silent grey boom.")[26] For those on the ground, the experience is reversed. Often, they cannot see the drones above but can hear a telltale buzz that is variously likened to a bee, mosquito, lawnmower, or generator. Because of the buzzing noise, Pashtuns in Pakistan call drones *machays* or "wasps," and Palestinians call them *zenana*, which means "buzz" but is also slang for a nagging wife.[27] An *Agence France-Pressse* journalist in Gaza tweeted that "the sound of what is apparently drones overhead has not stopped in hours. Sound like lawnmowers in the sky."[28] A study of drone strikes in Waziristan reported that "drones produce a monotonous buzz, almost like the sound of a generator, which together with the uncertainty that comes with the perpetual fear of missile strikes have had an immense psychological impact on the population. Particularly affected are young children who are said to be unable to sleep at night and cry due to the noise."[29] One Western aid worker likened the atmosphere to the aftermath of the September 11, 2001, attacks in the United States: "Do you remember 9/11? Do you remember what it felt like right after? I was in New York on 9/11. I remember people crying in the streets. People were afraid about what might happen next. People didn't know if there would be another attack. There was tension in the air. This is what it is like. It is a continuous tension, a feeling of continuous uneasiness. We are scared. You wake up with a start to every noise."[30]

Steven Coll, writing in *The New Yorker*, observes that "being attacked by a drone is not the same as being bombed by a jet. With drones, there is typically a much longer prelude to violence. Above North Waziristan, drones circled for hours, or even days, before striking. People below looked up to watch the machines, hovering at about twenty thousand feet, capable of unleashing fire at any moment, like dragon's breath. 'Drones may kill relatively few, but they terrify many more,' Malik Jalal, a tribal leader in North Waziristan, told me. 'They turned the people into psychiatric patients. The F-16s might be less accurate, but they come and go.'"[31]

Numerous commentators have remarked on the terrorizing effect of the drones' persistent presence, which can induce a state of paralysis in which people are afraid to venture into the open and instead hunker down indoors in what Gregoire Chamayou calls a state of "psychic imprisonment within a perimeter no longer defined by bars, barriers, and walls, but by the endless circling of flying watchtowers up above."[32] Doctors and psychiatrists have commented that those who live under the drones are increasingly showing many symptoms of post-traumatic stress, including breakdowns, nightmares, outbursts of anger and irritability, loss of appetite, sleeplessness, physical malaise, and unexplained physical pains.[33]

Western media reports generally have said little about the ways in which strikes have reshaped life for those who live under drones, particularly in the tribal areas of Pakistan. In addition to the pervasive sense of powerlessness, anticipatory anxiety, and dread mentioned above, journalists and human rights activists who have spent time in these areas report changes in customary practices in response to the threat of drone attacks, especially given the perception that drones are more likely to attack people when they gather in groups. These changes have affected political gatherings of elders, burial practices (already under pressure from the lack of intact bodies to bury), and the education of children: "Some community members shy away from gathering in groups, including important tribal dispute-resolution bodies,

Figure 2.4
A Pashtun war rug.
Source: Photo by Kevin Sudeith, warrug.com.

out of fear that they may attract the attention of drone operators. Some parents choose to keep their children home, and
children injured or traumatized by strikes have dropped out of
school. ... The strikes have undermined cultural and religious
practices related to burial, and made family members afraid to
attend funerals. In addition, families who lost loved ones or
their homes in drone strikes now struggle to support themselves."[34] Drones have even made their way into the fabric of
Pashtun rug designs (figure 2.4).[35]

Remixing Time and Space

Congressman Alan Grayson, a Democrat from Florida who has
been an outspoken critic of drones, emphasizes the distance and
impersonality of drone warfare: "A person sits in front of a

computer screen somewhere in the United States. He has never been to the target area, has never seen it from the ground, doesn't know anyone there, doesn't speak their language, isn't even familiar with the clothes that they wear. Based on what he sees on that computer screen, and whatever else he's got, he launches bombs from a drone aircraft flying in the sky 8,000 miles away. The bombs then kill people."[36] What Grayson describes is remote killing in every sense of the term: it is done by remote control, and it is spatially remote, culturally remote, and emotionally remote.

The history of military technology is one of increasingly remote killing as warriors have sought to kill one another with stronger doses of what Pink Floyd's Roger Waters calls "the bravery of being out of range."[37] Hand-to-hand combat of the kind memorialized in *The Iliad* gave way to the bow and arrow, which in turn gave way to the rifle, the machine gun, artillery, aerial bombardment, Cruise missiles, and intercontinental ballistic missiles. Throughout this centuries-long process in which military tactics and technology coevolved, combatants have sought ways of striking a blow from a sufficient distance that they themselves were immune to a reciprocal blow. Meanwhile, the increasing distance between warriors made killing more impersonal. A warrior in *The Iliad* looked into his enemy's eyes as he thrust a sword into his body, but World War II bomber crews found that their exploding bombs looked like fiery flowers 20,000 feet below and that it was hard to imagine the people they killed. Offshore ship crews that launch Cruise missiles may not even see their missiles explode. For drone operators, according to Gregoire Chamayou, "the act of killing is in effect reduced to positioning the pointer or arrow on little 'actionable images,' tiny figures that have taken the place of the old flesh-and-blood body of the enemy. ... One is never spattered by the enemy's blood. No doubt the absence of any physical soiling corresponds to less of a sense of moral soiling."[38]

The development of drones increases the distance from which a blow is struck to thousands of miles and triangulates a formerly dyadic relationship between combatants. In the arc of technological development described above, the weapon was moved farther away from its target, but the weapon's operator still had to be with the weapon to operate it. Along with landmines and improvised explosive devices (IEDs), drones have disarticulated the spatial relationship between weapon and warrior. In the case of drones, the weapon is proximate to the scene of destruction, but the person who controls it may be on the other side of the planet. What was formerly a tightly packed and spatially concentrated ensemble—weapon, weaponeer, and target—has been disarticulated.[39] In earlier wars, if a plane was shot down, the pilot died or was taken prisoner. The technology of remote control creates such invulnerability for drone pilots that, in the memorable locution of Richard Clarke, a former official of the National Security Council, "if the Predator gets shot down, the pilot goes home and fucks his wife."[40]

The respatializing dynamics here are profound and asymmetrical. The human targets of drone strikes feel trapped in the local, from which there may be no escape, but the targeters inhabit a space of free movement that has become stretched to global proportions. This is a little like the contemporary relationship between labor, which is trapped at the local level by lack of resources and by national boundaries, and globalized capital, which is free to move anywhere in the world with a few keystrokes. In the cases of both military conflict and global capitalism, the freedom to move on a global scale affords an important but not necessarily decisive advantage.

Drone warfare scrambles time and speed, too. With their ability to linger for hours, drones can track potential targets at a deliberate pace, but once a target has been confirmed, destruction is almost immediate. "In World War II, on many occasions target evaluation took months. Today it's single-digit minutes," said General David Deptula.[41] Once a target is in the cross-hairs drone technology speeds war up for the target while slowing it down for the targeter. Those targeted by drones have a fraction of a second when they realize (if they do) that an explosive is hurtling at them from the sky at hundreds of miles an hour. They reportedly hear a whooshing noise just before they are dismembered. The drone operators, on the other hand, may experience this climactic moment of kinetic action as if in slow motion, and it may be preceded by hours of patient surveillance, discussion of targeting options, and remote pursuit of the enemy across a variety of contexts and terrains. Because a drone can linger unseen in the sky for a long time, targeting decisions that were made in a more jagged, impulsive way in the past now can be slowed down and made more deliberately. And in war, the party that controls the tempo of hostilities has an advantage.

But it would be a mistake to say, as some commentators do, that drone warfare simply lies at the far end of a straight evolutionary line of technologies that increase the physical and emotional distance between combatants.[42] As quoted in the epigraph at the front of this book, the French philosopher Gregoire Chamayou says that "the drone upsets the available categories, to the point of rendering them inapplicable,"[43] and the category of space is no exception here. Drones enable an unprecedented distance between victims and executioners as someone on one side of the planet kills a person on the other side of the planet by pushing a "pickle" button, but at the same time the video feedback component of drone technology gives drone operators a sense of experiential immersion in their victims' deaths. The emotional force of this mediated experiential proximity is amplified by the requirement that drones linger after a strike to assess the damage and count the dead. Thus, the impersonality of remote killing is at least partially offset by what might be called remote intimacy. It is too reductive to say that the processes of respatialization in drone warfare simply distance drone operators from the battlefield, thus making killing easier. It is more accurate to say that they scramble relations of distance, making them simultaneously more elongated and more compressed in ways that are subjectively confusing and paradoxical. They make killing both easier and harder, creating a new psychological topography that we are struggling to understand.

We can make a similar argument about the spatial boundaries of the battlefield. Some commentators speak as if remote-control technology allows drone operators to act on the battlefield from outside, looking in like gods. American military and political leaders certainly speak as if the battlefield is over there in the Middle East, far away from what they increasingly call "the Homeland." But another way of thinking about this situation is to say that when a combatant acts from a place, then that place is by definition part of the battlefield. (The undergraduate students in a class I taught on the anthropology of war saw things this way: when asked if Taliban fighters would have the right to come to the United States and kill a drone operator as he drove home from work, most said yes.) If the battlefield exists wherever combatants are located, even if they are remote combatants, then drone operators have not entirely removed themselves from the battlefield but instead have globalized the battlefield, bringing experiential and organizational fragments of the battlefield inside the national boundaries of the homeland. Drone operators trap the targeted adversary within the local by acting from an unseen distance but at the same time enable shards of that faraway local battlefield to embed themselves in their own experience of the local. (The Hollywood movie *Good Kill* tries to evoke this reality through the device of repeatedly showing drone pilots' homes in Nevada as they would be seen from above by drones, as if they too were subject to the code of targeting imagery.) Thus, the flip side of globalization in this process of respatialization is that the clear boundary between the battlefield and civilian space in the U.S. homeland is in danger of erasure. Drone strikes on houses and cars in Pakistan, Yemen, and Somalia—countries with which the United States is not formally at war—tend to undermine the distinction between civilian spaces and military targets and the notion of a separate battlefield that has hitherto been enshrined by the international laws of war. At the same time, the removal of military pilots from bases abroad to bases within the United States that are physically distant but experientially proximate to the killing zones accomplishes a parallel erasure of the distinction between combat and noncombat spaces.

Many drone pilots report experiencing this embrittlement of the distinction between the inside and outside of the battlefield particularly in relation to their family lives. This is not (as for foreign militants targeted by drones) because the U.S. combatant's family may be struck at any time by the enemy but because the experience of battle in a faraway local cannot be quarantined from the U.S. combatant's domestic life.

Customary military practice in wartime has been to remove combatants from their families and segregate them on military bases with their fellow combatants, who become their new family. (It is no accident that fellow soldiers have often called one another brothers.) This separation of combatants from the symbolically feminine space of domestic life keeps combatants hard and helps maintain discipline, a sense of esprit de corps, and continuous immersion in the project of war. For drone operators, this separation of warriors from their families is undermined by the rhythms of deployment at home, and for soldiers deployed to the Middle East, it is undermined by the ubiquity of Skype—another example of the power of virtual communications technology to play havoc with the conventional boundary between combat and domestic spaces.[44] For drone operators, now turned into commuters who move back and forth between their families and the immersive virtual killing zones of their pods, the proximity of work to family creates confusion between the battlefront and the homefront, which are separated only by the temporal boundaries of shiftwork. "A killer in the morning and a father in the evening, a daily switch from the 'peace ego' to the 'war ego,'" as Chamayou puts it.[45] In George Brant's play *Grounded*, the husband of the fictional drone pilot makes her French toast for breakfast before she sets off for her "first day on the job, the war, whatever." She says:

Home will be training too
Getting used to the routine
Driving to war like it's shift work
Like I'm punching the clock
Used to transition home once a year
Now it's once a day.

She adds:

It would be a different book
The Odyssey
If Odysseus came home every day.[46]

Stars and Stripes quotes a drone operator as follows:

You've just been on a combat mission and half an hour later your spouse is mad at you because you're late to soccer practice.

Friction at home can stem from just that simple question upon walking through the door: "How was your day?"[47]

The difficulty in navigating the contradictions between these two spaces is a recurrent theme in Matt Martin's memoir. He describes his daily schedule as follows: "commute to work in rush-hour traffic, slip into a seat in front of a bank of computers, 'fly' a warplane to shoot missiles at an enemy thousands of miles away, and then pick up the kids from school or a gallon of milk at the grocery store." He finds it jarring to be pulled over for speeding when he is fighting a war: "What was the matter with this policeman? Didn't he realize I was at war? At any other point in history, it would have been inconceivable that a combat pilot could take time out from fighting to have a leisurely breakfast with his wife and then get a speeding ticket on the way to work. Another of those strange juxtapositions of alternate lives from two vastly different lives." Later he says, "it was enough to make a Predator pilot schizophrenic, what with fighting two wars simultaneously 1,500 miles apart and balancing them with a wife and kids, if he had them, paying the bills, and calling the plumber because the toilet was stopped up. It didn't get much more surreal than that."[48]

Martin reports that for a long time he could not talk to his wife about what he did in the pod, especially the wrenching

details and the accompanying sense of anguish when he accidentally killed two children and an old man. Instead, he com-
partmentalized. He was partly relieved when he was sent to Iraq,
away from his wife, as part of a ground crew on a military base
there. Another drone operator reports, "I also spent some time
operating drones in Afghanistan, and when you finished a day's
work there, you were still in a combat zone, on a military base
surrounded by people walking around with guns. You can
decompress more easily by chatting to them and comparing
experiences, whereas at Creech I would just go home to the wife
and kids."[49]

The psychological torquing created by the oscillation between
the worlds of family domesticity and onscreen killing is the central theme in the two best works of fiction about drone operators—George Brant's play *Grounded* and Andrew Niccol's film
Good Kill,[50] which stars Ethan Hawke as a drone operator. In both
works, the central drama is the gradual psychic disintegration of
a drone pilot, a woman in *Grounded* and a man in *Good Kill*,
brought on in large part by the corrosive effect of their work on
their domestic life. In *Good Kill*, the contradiction leads, seem-
ingly inexorably, to divorce.

Remixing Valor

"I don't know what I'm doing, but it's not flying," says the fictional drone operator in the Hollywood movie *Good Kill*. There
are understandable ambiguities about drones and the people
who operate them. Are they pilots if they are not in an airborne
aircraft? Are they combatants if they are thousands of miles
away from the battlefield? Do they deserve combat medals?
Although Predator and Reaper aircraft are called drones by the

media and the proverbial man on the street, the U.S. Air Force calls them remotely piloted aircraft (RPAs), implying that the people who control them are pilots, even if they are not inside the planes. The drone operator Bill "Sweet" Tart fines his subordinates $5 every time they use the word *drone*: "The word *drone* is a negative with respect to the skill and effort that the men and women individually put into flying and executing a mission," he says. "A drone, whether you're talking about a drone bee that does no work, or a drone that uses artificial intelligence and does its mission without any human input, is not what we're talking about here."[51] The air force insists that the people who control drones are pilots, makes them wear flight suits on duty, and gives them flight pay, but their uniform wings are not the same as the uniform wings of pilots of manned planes. The badge for those who fly unmanned aerial vehicles (UAVs) features a lightning bolt to symbolize the satellite signal that controls a drone (figure 2.5).[52] Moreover, there is a strong symbolic hierarchy within the Air Force with fighter and bomber pilots at the top, followed by tanker and cargo pilots, then drone operators at the bottom. Drone operators sometimes complain that they find it hard to get promoted no matter how well they do their jobs,[53] and pilots of manned planes have been known to deride them as

Figure 2.5
Unmanned aerial vehicle operator badge.

the "Chair Force" and "Chairborne Rangers."[54] In a *New Yorker* article on drones, Jane Mayer puts "flown" in scare quotes and remarks on the incongruity of the fact that drone operators are said to wear flight suits. As for whether they are combatants, she calls them "cubicle warriors [who] can drive home to have dinner with their families."[55]

There is also the issue of medals. How should the state recognize contributions on the battlefield that, however important, involve no risk of bodily harm? Some drone operators complained that drone crews spent 630 hours searching for the al Qaeda leader Abu Musab al-Zarqawi and received a thank-you note after they found him, but that the F-16 pilot who killed him, "who faced no real threat from the lightly armed insurgents on the ground, was awarded the Distinguished Flying Cross, the same honor bestowed on Charles Lindbergh for the first solo flight across the Atlantic Ocean."[56]

In February 2013, the Pentagon announced a new Distinguished Warfare Medal (figure 2.6) that ranks higher than the Bronze Star but lower than the Silver Star and recognizes drone operators for, in Secretary of Defense Leon Panetta's words, "extraordinary achievements that directly impact on combat operations, but that do not involve acts of valor or physical risk that combat entails."[57] Not atypical of the outraged reactions among many military veterans was a satirical article on *Duffelblog*, a humorous military news website. Describing a fictitious Major Beasley who won the new medal, it said, "The medal will be in addition to the multiple Purple Hearts he has already been awarded for Carpal Tunnel Syndrome endured during the same event. He was also honored for a sprained ankle he received tripping over an extension cord while leaving his workstation." The article added that "Beasley's wife and family have also released a

Figure 2.6
Distinguished Warfare Medal, February to April 2013.

statement saying how proud they were, and how they had eventually forgiven him for missing family Scrabble night during the fighting."[58] Dos Gringos, two military aviators who record country rock music, distributed a song called "Predator Eulogy," which included this verse:

They shot down the Predator
I wonder how that feels
For that operator who has lost his set of wheels
It must be so defenseless
Like clubbing baby seals.[59]

Two months after the introduction of the Distinguished Warfare
Medal for drone crews, the new Secretary of Defense, Chuck
Hagel, canceled it.[60]

Casting aspersions on the valor of soldiers who are not on
the frontlines is not a new practice, especially if they operate
new technologies whose place is not yet settled. In 1139, Pope
Innocent II's Second Lateran Council condemned the crossbow
because it "could be used from a distance, thus enabling a man
to strike without the risk of himself being struck."[61] When submarines were first developed, their crews were accused of cowardice because they sank ships whose crews had no idea the
submarines were nearby.[62] When guns were first introduced,
they were condemned as a weapon for cowards.[63] Despite
their valorization in Clint Eastwood's movie *American Sniper*,
snipers have also been accused of cowardice. But drones seem
to attract particularly strong opprobrium, from both inside and
outside the military. Casting aspersions on the manhood of
drone operators, Gregoire Chamayou suggests that calling
drone flights "unmanned" is appropriate in a double sense.[64]
Glenn Greenwald says, "Whatever one thinks of the justifiabil-
ity of drone attacks, it's one of the least 'brave' or courageous
modes of warfare ever invented. It's one thing to call it just,
but to pretend it's 'brave' is Orwellian in the extreme. Indeed,
the whole point of it is to allow large numbers of human
beings to be killed without the slightest physical risk to those
doing the killing. Killing while sheltering yourself from all risk
is the definitional opposite of bravery."[65] Jonathan Schell asks
if we can even call the asymmetrical operations of drones
"war." Discussing Barack Obama's claim in 2011 that the War
Powers Act did not apply to aerial intervention in Libya
because no Americans would die, Schell says,

War is only war, it seems, when Americans are dying—when we die. When only they—the Libyans—die, it is something else for which there is as yet apparently no name. ... In our day, it is indeed possible for some countries, for the first time in history, to wage war without receiving a scratch in return. ... The epitome of this new warfare is the Predator drone, which has become an emblem of the Obama administration.[66]

"Absent fear, war cannot be called war," says commentator Steve Featherstone. "A better name for it would be target practice."[67]

This is why drone strikes are widely perceived in the Middle East as cowardly, and it is why Jody Williams, the Nobel Peace Prize winner, said, "when you sit in Nevada and kill someone 7,000 miles away, I think there's something unethical or immoral about that."[68] Sir Brian Burridge, a former Royal Air Force air chief marshal in Iraq, has called drone warfare "virtueless war,"[69] and Shane Riza, an air force pilot and instructor, writes that he is "concerned the last dying embers of a warrior culture are about to be snuffed out in favor of a system of warfare that just might destroy war's very meaning."[70] Andy Exum, a former U.S. Army officer who is now an expert on counterinsurgency at the Center for a New American Security, says, "As a military person, I put myself in the shoes of someone in FATA [Pakistan's federally administered tribal area] and there's something about pilotless drones that doesn't strike me as an honorable way of warfare." Invoking the warrior ideals of *The Iliad*, he adds, "As a classics major, I have a classical sense of what it means to be a warrior."[71]

Drone warfare is continuous with a long tradition of colonial war-fighting technologies—going back at least to the machine guns that nineteenth-century British and French colonial soldiers used to mow down spear-carrying Africans—that ensure that "natives" die, in an unfair fight, in considerably larger numbers than colonial soldiers.[72] In fact, the drone is not envisaged as a weapon that would be used in fighting evenly matched adversaries such as the Russians or the Chinese, who would quickly shoot a lumbering Predator out of the sky. It is designed to save American lives in highly asymmetrical postcolonial counterinsurgency operations—what some in the military refer to as "small wars." The asymmetry of drone operations thus derives not just from the technology itself and the ways that it absents one party to the combat from the scene of combat, but from the neocolonial context of combat. In neocolonial counterinsurgency contexts, there is invariably a massive asymmetry in casualties when outnumbered occupying forces use superior technology to subdue indigenous populations.

In his book *On Suicide Bombing*, the anthropologist Talal Asad observes that in the era of drones, American "soldiers need no longer go to war expecting to die, but only to kill. In itself, this destabilizes the conventional understanding of war as an activity in which human dying and killing are exchanged."[73] Asad sees the honorable drama at the core of combat as one in which contending soldiers meet to wager their bodies for a cause. The willingness to forfeit one's own life lends some meaning to war, and the mutual availability of the bodies of combatants on both sides for injury and death establishes an honorable reciprocity between enemies and affords soldiers opportunities for the courage that is war's defining virtue. Seen through Asad's lens, the absence of this reciprocity of bodily exposure in drone warfare makes it dishonorable. It also suggests that there is a perverse parallelism, or mirror imaging, between drone warfare and the tactic of suicide bombing that some insurgents have adopted. Drone operators and suicide bombers, either by preemptively destroying their bodies or by absenting their bodies, deprive their adversaries of the opportunity to capture or kill them, thus undermining the structural reciprocity that conventionally, or at least ideally, defines war. This is partly why the victims of both drone strikes and suicide bombings brand these modes of killing as terrorist. In contemporary war zones where people on the ground are blown to pieces either by unseen American drones in the sky or unsuspected suicide bombers in their midst, what we conventionally understand by the word *war* is being torn apart from above and below by American technology and insurgent tactics.

Seen from one perspective, the asymmetry between American immunity and the casualties of the other in the killing fields is a comfortable development. General Charles Dunlap says, "I don't feel better if a lot of Americans die in the effort. I'm ok with all of *them* [that is, the other side] dying."[74] On the other hand, Jane Mayer quotes a U.S. veteran of the Iraq War who says, "There's something important about putting your own sons and daughters at risk when you choose to wage war as a nation. We risk losing that flesh-and-blood investment if we go too far down this road."[75] Although we can understand why political and military leaders would be drawn toward a model of war in which only the other side dies, this veteran's claim is important and is further explored in chapter 5.

3 Remote Intimacy

They are inches apart. But Sergeant Jessica Aldridge is also eight thousand miles away, ten thousand feet in the air, and she feels so near to the figures on the ground below her that she might reach down and pick them up like dolls.

—Ron Childress, *And West Is West*[1]

We live in a world in which digital activities—video games, text messages, Facebook friendships, Instagram, X-box, selfies, online sex, talking to Siri, and interactive multiuser games—are focal points of Americans' social and emotional lives. This is a world in which *Her*, a 2013 Hollywood movie (directed by Spike Jonze) about a man who falls in love with his computer, became a box-office hit. "Life on screen" is increasingly absorbing, and in the words of James William Gibson, the science fiction writer credited with coining the term *cyberspace*, "everyone who works with computers seems to develop an intuitive faith that there's some kind of actual space behind the screen."[2] The emergence of drone warfare, organized around screen killing,[3] is part of this mass emotional and perceptual migration to screen worlds. In the words of actor Ethan Hawke, reflecting on the character he plays in the 2014 film *Good Kill*, "It's not a huge jump from what's happening to these pilots to what's happening to all of us. ... More and more of our intimacy, what used to be real and tangible, is now automated, is now from a distance."[4]

In her books *Life on Screen* and *Alone Together*, the sociologist Sherry Turkle writes compellingly about the immersive, emotionally compelling, and addictive qualities of screen worlds. Talking about "the seductions of the interface" and "computer holding power," she describes children who are disappointed when they see seals in a zoo because they do not behave as colorfully as the seals do in heavily edited television shows: "we are moving toward a culture of simulation in which people are increasingly comfortable with substituting representations of reality for the real. We use a Macintosh-style 'desktop' as well as one on four legs. We join virtual communities that exist only among people communicating on computer networks as well as communities in which we are physically present. We come to question simple distinctions between real and artificial. In what sense should one consider a screen desktop any less real than any other?"[5]

Remote Watching

An uneasy combination of physical remoteness and vivid mediated presence lies at the core of the drone operator's experience. "One of the paradoxes of drones is that, even as they increase the distance to the target, they also create proximity," says the journalist Nicola Abé.[6] In the words of playwright George Brant, that target is "twelve hours ahead but only 1.2 seconds away."[7] The operator sits in a trailer on the ground in what looks like a cockpit, "flying" a plane that is thousands of miles away. He or she stares for hours at a time at screen footage of the ground below the plane but is cut off from the view outside the trailer, and is drawn into the world onscreen through a process of mental slippage. Like blindfolds, hoods, darkened cells, or isolation tanks, the air-conditioned trailers seal off drone operators from the nearby physical environment and make them suggestible to other cues. The installation of a cockpit environment inside the trailer facilitates the illusion of being inside the drone. Sometimes the lack of authentic embodied experience can be a problem: one pilot crashed her drone because she did not realize that it was flying upside down, and another did not realize until too late that he had put his drone into a spin.[8] On the other hand, there are many stories of pilots who react as if they are inside their drones. Peter Singer says that "one officer recalls that the action felt so intense that one time, when his drone thousands of miles away was about to crash, he instinctively reached for the ejection seat."[9] Drone operator Matt Martin tells a similar story about an incident when his drone was on a collision course with an F-18: "I was so into the moment that every muscle in my body tensed for the impact. I leaned into the turn with adrenaline pumping. I couldn't have been more involved had I actually been inside the plane. ... For just a moment, I reverted to survival training instinct and thought about ejecting." Throughout his book, Martin speaks of drones as if their controllers are inside them: "Watchdog Four-Six was airborne with Bobby and Rexford in the cockpit"; "Bobby was patrolling above only seconds away"; "One afternoon when she and Bobby were cruising over the town of Balad"; and "I didn't have to worry about some pesky MiG showing up to shoot *me* down" [emphasis added].[10]

Drone operators spend more of their waking hours staring at video footage of ground terrain thousands of miles away than looking at the landscape where they live. Even if they complain how boring the work usually is, they become cognitively and psychologically immersed in the screen world. One drone operator who did the night shift over Afghanistan, scanning the terrain with an infrared camera, reports that he began to dream in infrared.[11] Operators' descriptions of hours a day spent in "overwatch" evince a sense of being simultaneously remote and present, and the operators sometimes speak of themselves as if they are displaced into the drone and watching from a great height, not from a trailer on the ground.[12]

This space of all-seeing power, where one sees without being seen, is often described metaphorically as the space of the gods. Martin says of insurgents that "the poor bastards never once considered looking up, *way* up, from which heights Predator crews observed their every move, where they went and who they met with. … That an eye from the inner edge of space might be watching was too far-fetched for them to imagine." He also says that "not once had I observed a single insurgent with a tube or visiting a weapons cache who so much as looked up to scan the sky. It was almost like they were ants down there, predictable in their behavior to some degree of mathematical probability, no more aware of Predator's presence than they were of the Almighty watching them." In this situation, Martin says, "Sometimes I felt like God hurling thunderbolts from afar," and "I truly felt a bit like an omnipotent god with a god's seat above it all."[13] Such turns of phrase are a common trope in commentary about drones.

As the drones gaze unblinkingly from above, there can be voyeuristic pleasure in watching the Other. In fact, it is hard to imagine a more voyeuristic technology than the drone. Martin, who describes himself as "a voyeur in the sky," likens his work to "watching some reality TV program that went on endlessly."[14] Photography critic Liz Wells defines *voyeurism*, which as a fetish often involves an obsessive desire to gaze at the hidden and the secret, as "a mode of looking related to the exercise of power in which a body becomes a spectacle for someone else's pleasure."[15] The act of voyeurism, which establishes the dominance of the watcher over the watched, connects while reinforcing distance. Sometimes, in keeping with the colloquial understanding of voyeurism, it involves watching sex acts. The journalist Nicola Abé gives an example of this: "When it got dark, Bryant switched to the infrared camera. Many Afghans sleep on the roof in the summer, because of the heat. 'I saw them having sex with their wives. It's two infrared spots becoming one.'"[16] And Martin describes an incident he watched in which an Iraqi "walked up to where a small grayish mule was browsing in the field. He stopped to look around before he looped a rope around the animal's neck and tied it to a shrub. The guy lifted his man-dress, approached the mule from the rear, and, without further foreplay, began to service it. I called Bobby over to take a look. ... My guy down there had no idea he was starring in his own video, the Iraqi version of *Debbie Does Dallas* with a more perverted twist."[17]

Voyeuristic pleasure can also be taken in watching people die. "It can be hard for people to take their eyes off it. If it's an operation to hit some well-known terrorist, you'll see people crowding around the screens," reports a British drone operator.[18] Peter Singer describes a genre of video clip he calls "war porn": "Clips of particularly interesting combat footage, such as an insurgent blown up by a UAV, are forwarded to friends, family, and colleagues with titles like 'Watch this!' ... Comments and jokes are attached, and some are even set to music. A typical example was a clip of people's bodies being blown up into the air by a Preda-tor strike set to Sugar Ray's song, 'I just want to Fly.'"[19]

Speaking to the appeal of what he calls "Predator Porn" to rank and file soldiers, David Kilcullen, an expert on counterin-surgency who advised the Pentagon, says:

In a counterinsurgency environment, you almost never see the enemy, you certainly never see them doing anything that's bad. Whenever you do encounter them, they're trying to hide amongst the population. That's incredibly frustrating for people. So when people can see a bad guy carrying a weapon, acting like a bad guy, getting blown up, it's enor-mously satisfying for some Humvee guy who goes out and spends all day driving around Baghdad getting shot at. He comes back and he fires up some gun camera footage from an Apache or drone strike footage and says, "Well at least we got some guy today."[20]

Remote Narrativization

This distanced voyeurism is counterbalanced by a sense of immersive intimacy when drones patrol the same terrain over and over and operators come to feel they know the people below. A profile of drone operator Bill "Sweet" Tart in the *Huff-ington Post* observes that "Peering through cameras and sensors from his computer station thousands of miles away, he absorbs the details of daily life in the villages below. He develops an eerie intimacy with his targets. Which house these kids belong to. When that mom goes out to market. Who visits and why. ... 'You can start figuring out who is associated with who. Who is a stranger, who is it that's visiting this house?'"[21] Another drone operator says, "You get a sense of daily life. I've been on the same shift for a month and you learn the patterns. Like, I'll know at 5 a.m. this guy is gonna go outside and take a shit. I've seen a lot of dudes take shits." Reflecting on the way that his job helps him understand a radically different way of life, the same man reports, "Another time we followed this guy outside his house for half an hour, and all he did was go scoop water from a stream. Seeing that just made it sink in—how we live worlds apart."[22]

Matt Martin tells a story in his memoir that illustrates nicely the "eerie intimacy" that operators can experience with those they watch. He was circling over "a man equipped with a broom and a rusty shovel with a broken handle." The man "began working on one of those dilapidated roads that branched off the thoroughfare from the mosque." Martin kept watching because "someone digging in a road was always cause for suspicion. He could just as well be a saboteur planting a bomb." The man "straightened and arched his back, stretching. He yawned. He wiped sweat from his brow. He appeared to be in his thirties and, best I could tell, had very little spare flesh on his rib cage. I finally concluded that he was exactly what he seemed to be—some ordinary Iraqi citizen clearing debris off the road and filling in potholes, tidying up in hopes that the unused mosque might be revitalized." Then, with Martin still watching, a U.S. convoy drove by the man. The heavy vehicles "cracked the macadam surface and reopened potholes, destroying all the progress the worker had made." Martin tells the reader with sympathy that the man "stood with his broom and broken shovel by the side of the road and gazed dejectedly at the armored convoy as it drove off and over the horizon. His shoulders slumped and his chin dropped on his chest. I recognized the posture even from ten thousand feet above him. Defeat. Poor bastard."[23]

Martin's narrative offers an example of remote narrativization. As drone operators watch people on the other side of the world from thousands of feet above, they create mental stories that help make sense of the people they watch. In the process, they can make interpretive leaps, fill in informational gaps, and provide framing moral judgments as they integrate shards of visual information and turn pixelated figures into personalities. "You can't see their faces," says the fictional drone pilot in George Brant's play, *Grounded*, "but you don't need to[;] your mind fills them in."[24] In the story above, by watching what amounts to a fragmentary silent film without dialogue, Martin concludes that the man he is watching is not an insurgent but a decent, hard-working man who is trying to revitalize his local mosque and that his reaction to the American soldiers who destroy his work is dejection, not rage. At another point, Martin sees "a one-story flat-roofed building with a small vendor's sign hanging over the door. I couldn't quite make out the lettering, which would not have helped me much anyhow because it was in Arabic. I guessed it was probably a law office."[25] But how could he possibly know it was a law office? Such moments of overnar-rativization are often based on unconscious cultural assumptions and on seeing things that may or may not be there, but they can determine whether people live or die. As is shown in the next chapter, drone operators have on numerous occasions killed people they "knew" to be insurgents only to find out later that they had killed innocent civilians. In the words of journalist Andrew Cockburn, "there is a recurrent pattern in which people become transfixed by what is on the screen, especially when the screen—with a resolution equal to the legal definition of blind-ness for drivers—is representing people and events thousands of miles and several continents away."[26]

In such processes of narrative infilling, there is a tendency to think that one knows more than one does. The potential for calamity here is dramatized in a story told by David Cloud, an investigative journalist for the *Los Angeles Times*. Working from interviews and transcripts of drone operators' conversations pried loose by a Freedom of Information Act request, Cloud reconstructed an attack in Afghanistan in February 2010 that, according to Afghan villagers, killed twenty-three civilians, including three- and four-year-old children. (The United States said that fifteen or sixteen had been killed but conceded they were civilians.) A group of Afghan civilians had set out before dawn on a long trip, driving in a convoy so that if one driver's car broke down, the others could assist. The group included shopkeepers buying supplies, students returning to school, and villagers visiting relatives or seeking medical treatment. A U.S. Army Special Forces team was active in the area nearby, and radio intercepts suggested that insurgents might be planning to attack the team. A Predator drone was overhead, watching unseen. It tracked the Afghans on the ground for four and a half hours and gathered information to confirm that they were insurgents. Throughout, the drone's crew conferred with the leader of the U.S. team on the ground and a team of "screeners" at a military base in Florida. The screeners were trained in the analysis of video imagery, but according to the *Los Angeles Times*, "even with the advanced camera on the Predator, the images were fuzzy and small objects were difficult to identify. Sometimes the video feed was interrupted briefly."[27]

When a drone operator in Nevada first saw members of the Afghan convoy, they seemed to be kneeling on blankets. Assuming that kneeling signifies praying and that praying is a habit of insurgents, the drone operator began connecting informational dots that should not have been connected. "'They're praying,' he said. 'This is definitely it, this is their force. … Praying? I mean, seriously, that's what they do.'" When one driver on the ground flashed his headlights at another, the pilot of an AC-130 above radioed that the two vehicles were, as one would expect of insurgents, "signaling." Assumptions that the people were all "military age males" also excited suspicion. One person involved in the incident later said, "We all had it in our head, 'Hey, why do you have 20 military age males at 5 a.m. collecting each other?' There can be only one reason."[28]

When the drone pilot and his camera operator conferred about what the infrared camera was picking up in the darkness, the pilot said, "See if you can zoom in on that guy. Is that a rifle?" The camera operator said, "Maybe just a warm spot from where he was sitting. Can't really tell right now, but it does look like an object." Fifteen minutes later, the camera operator returned to the topic: "Yeah, I think that dude had a rifle." "I do too," said the pilot.[29] Even if they did actually see a rifle, the use of a rifle as a metonym for an insurgent is problematic because Afghan civilians are often as heavily armed as NRA loyalists.

Shortly after the camera operator observed that the "truck would make a beautiful target," one of the screeners in Florida said that he had seen a child in the group. "Bullshit! Where?" said the camera operator. "I don't think they have kids out at this hour." The pilot complained, "Why are they so quick to call kids but not to call a rifle?" The camera operator said that there might be a teenager in the group but no one who was short enough to be a child. Nearly three hours after the convoy was first spotted, the "screeners" in Florida estimated "21 MAMS [military-aged males], no females, and two possible children." The drone camera operator responded, "Not toddlers. Something more toward adolescents or teens." The pilot agreed, and a soldier on the ground radioed, "Twelve or 13 years old with a weapon is just as dangerous." Although the vehicles in the convoy were now moving away from U.S. soldiers on the ground, they were thought to be involved in a flanking maneuver, and an attack was called in. "The Predator crew in Nevada was exultant, watching men they assumed were enemy fighters trying to help the injured," wrote the *Los Angeles Times* reporter. "I forget, how do you treat a sucking chest wound," crowed one member of the drone team.[30]

Doing damage assessment, the drone operators soon realized that they were watching women and children in the wreckage. In an abrupt 180-degree turn on the reliability of their aerial observations, one member of the crew said, "No way to tell, man." The camera operator agreed: "No way to tell from here."[31]

Major General James O. Poss oversaw an air force investigation of the incident and conceded that "technology can occasionally give you a false sense of security that you can see everything, that you can hear everything, that you know everything."[32] The phrasing suggests that technology itself might be to blame, but of course the fault lies in the interaction between the limitations of the technology and processes of narrative infilling and remote individualization. In these processes, after a frame has been put in place, ambiguous information is interpreted within that frame, informational gaps are ignored, and moral judgments are rendered. In the story just told, the frame is that the people on the ground are insurgents. Praying, flashing lights, and proximity to U.S. troops confirm this, and the frame becomes so powerful that visual evidence that children are present is discounted. The interpretation is strengthened by incorrect cultural judgments that are rendered from afar by people who have not spent time with the people whose behavior they are evaluating: they think that only insurgents pray and carry guns and that children in Afghanistan are never on the road before dawn. Throughout the communicative process, participants do not check and question each other but instead confirm each other's cognitive misinterpretations. A Special Forces sergeant who viewed the video later said, "Someone was saying when the vehicles stopped, the [passengers] were praying. ... When I looked at the video, they could also have been taking a piss. Whoever was viewing the video real-time, maybe they needed a little more tactical experience. It needs to be someone who knows the culture of the people. If I can say anything, they just need to be familiar with what they are looking at."[33]

The excerpts from the transcripts in the *Los Angeles Times* article reveal a palpable hunger to attack—what a United Nations official who read the transcript dryly called "a predisposition to engage in kinetic activity"[34]—that is at odds with the dispassionate and careful evaluation of potential targets that drone operators usually ascribe to themselves in interviews with journalists. Major General Timothy McHale, who also investigated the incident, came away with the impression that members of the Predator crew "were out to employ weapons no matter what."[35]

Screen Killing

Many critics of drone warfare say that it is psychologically easy for drone operators to kill because screen killing is just like playing a video game. In a report titled *Convenient Killing: Armed Drones and the 'PlayStation' Mentality*, the Fellowship of Reconciliation, an interfaith Christian organization, says that "operators, rather than seeing human beings, perceive mere blips on a screen."[36] The report quotes Philip Alston and Hina Shamsi, whose column for *The Guardian* newspaper condemned the "'PlayStation mentality' that surrounds drone killings. Young military personnel raised on a diet of video games now kill real people remotely using joysticks."[37] In a similar vein, Pratap Chatterjee disdains what he calls "desktop killing,"[38] and anthropologist Jeffrey Sluka writes that for drone operators "killing is just a matter of entering a screen command; to the drone pilot, it is like pushing 'Ctr-Alt-Del' and the target dies."[39] A report published by the International Committee of the Red Cross states that "advanced technologies which permit killing at a distance or on the computer screen prevent the activation of neuro-psychological mechanisms which render the act of killing difficult."[40] These critiques suggest that screen killing is too easy because it looks aesthetically like playing a video game and because, in the words of the Fellowship of Reconciliation report, "the geographical ... distance between the drone operator and the target lowers the threshold in regard to launching an attack."[41]

If we compare the experience of playing a video game and operating a drone, there are important differences as well as similarities. The geographer Derek Gregory points out that "immersion is video games is discontinuous—levels are restarted, situations re-set, games paused—and while there are different intensities of involvement during a UAV mission and shifts change in the course of a patrol, immersion in the live video feeds is intrinsically continuous." Gregory also points out that, unlike drone camera feeds, video games "show stylized landscapes prowled solely by 'insurgents' or 'terrorists' whose cartoonish appearance makes them instantly recognizable."[42]

But the critics are making an argument about distance as well as aesthetics, saying that it is disturbingly easy to kill someone if they are 8,000 miles away. These arguments are indebted to the book *On Killing* by Dave Grossman, a former U.S. Army lieutenant colonel who is credited with founding the field of "killology." Grossman reports that many World War II infantrymen never fired their rifles, even under attack, and argues that "at close range the resistance to killing an opponent is tremendous. When one looks an angry opponent in the eye, and knows that he is young or old, scared or angry, it is not possible to deny that the individual about to be killed is much like oneself."[43] Grossman's book includes a much-cited graph that plots a linear relationship between ease of killing and the physical distance separating the killer and the killed, ranging from "sexual distance" to "bomber range." The people who find it easiest to kill, according to this model, are those who are farthest away from their victims and therefore do not confront the humanity of those they kill. These include bomber crews and nuclear missile launch control officers.

But the remote intimacy of drone operations plays havoc with Grossman's model and these critics exaggerate and misconstrue the psychological ease with which drone operators kill. Drones take the straight line in Grossman's graph and twist it into a Mobius strip where beginning and end, although still separate, cross. Physically, the drone operator is as far from his victims as the intercontinental ballistic missile (ICBM) officer in an underground bunker in Montana is. But the drone operator and the ICBM officer are worlds apart perceptually: ICBM officers would never see their victims, but drone operators see them on screen before and after killing them. Drone operators may consummate hours or days of intimate watching with killing in a way that makes the violence in some ways more psychologically proximate than that of other soldiers, who are physically closer to the enemy but may get only a brief glimpse of, or never see at all, the people they kill. A journalistic profile of one drone operator remarks that "he observed people for weeks, including Taliban fighters hiding weapons, and people who were on lists because the military, the intelligence agencies or local informants knew something about them. 'I got to know them. Until someone higher up in the chain of command gave me the order to shoot.' He felt remorse because of the children, whose fathers he was taking away. 'They were good daddies,' he says."[44]

Grossman's argument also assumes that there is something about physical proximity and especially a clear line of sight between people that excites an instinctive empathy for another's pain and a reluctance to kill. There is surely some truth to this, but as Elaine Scarry argues in her book *The Body in Pain*, no matter how close people might be, there is an "unsharability" to another's pain. "The events happening within the interior of that person's body may seem to have the remote character of some deep subterranean fact," Scarry says: "When one speaks about 'one's own physical pain,' and about 'another person's physical pain,' one might almost appear to be speaking about two wholly distinct orders of events."[45] We do not apprehend another's pain experientially, as we do our own, but instead by deciphering signs. Cries, grunts, facial expressions, and so on can become cues for empathy with the suffering of another. Although drone operators do not hear cries or see facial expressions, they are exposed to other signs of pain that bomber pilots probably are spared. The former drone operator Brandon Bryant touched on this in an interview about the death of an insurgent he killed:

The guy in the back [one of three insurgents being tracked] hears the sonic boom when it reaches him, and he runs forward. We're told to get the two guys in the back, worry about the guy in the front later. ... The guy in the back runs forward between the other two and we strike all three of them. The guy that was running forward. ... when the smoke clears, there's a crater there. He's missing his right leg and I watched this guy bleed out. And it's clear enough that I watch him. He's grabbing his leg [grabs own leg on camera] and he's rolling. I can almost see the agony on this guy's face. And eventually this guy becomes the same color as the ground that he bled upon.

Q: So he loses his heat. You watch him die.
A [uncomfortable laugh]: Yeah! You know how people say that drone strikes are like mortar attacks or artillery. Well, artillery doesn't see this, artillery doesn't see the result of their actions. It's really more intimate for us because we see everything. We see the before, action, and after. And so I watched this guy, I watched him bleed out. I watched the result of [long pause], I guess collectively it was our action, but ultimately I'm the responsible one who guided the missile in.[46]

One could say that watching a distant figure change color on an infrared screen is less affecting than seeing blood spurt from a leg wound and hearing a person's cries and moans. But what Bryant saw was enough to make him instinctively grab his own leg in sympathy during the interview, and this image was vivid enough that it haunted him, with the draining of color on screen becoming a powerful visual metaphor of death. "How many women and children have you seen incinerated by a Hellfire missile? How many men have you seen crawl across a field, trying to make it to the nearest compound for help while bleeding out from severed legs?" asks Heather Linebaugh, a former drone imagery analyst: "When you are exposed to it over and over again it becomes like a small video, embedded in your head, forever on repeat, causing psychological pain and suffering that many people will hopefully never experience."[47] "You see a lot of detail," says the commander of a drone squadron. "So we feel it, maybe not to the same degree [as] if we were actually there, but it affects us. ... When you let a missile go, you know that's real life—there's no reset button."[48]

The work of drone operators is marked by what anthropologist Antonius Robben calls "ambivalence of enemies as both human and virtual."[49] There is an oscillation in drone operator Matt Martin's accounts of people he killed. At one point, he says "It was almost like watching an NFL game on TV with its tiny figures on the screen compared to being down there on the field in the mud and the blood in the rain." After his first experience of virtual combat, he says, "It would take some time for the reality of what happened so far away to sink in, for 'real' to become *real*." But only a few pages later, he says that "those who would call this a Nintendo game had never sat in my seat. Those were real people down there."[50] And when Martin accidentally kills two children, there is no sense that they are just "blips on a screen" or that this is "PlayStation warfare." Instead, screen killing triggers a cascade of associative memories and vivid images. Just after he had released a missile,

two kids on a bicycle unexpectedly appeared on the screen approaching the truck and the insurgents. Both were boys. One appeared to be about eleven, the other—possibly a younger brother—was balanced on the handlebars. Tooling along on a summer day laughing and talking."

"Oh God! Not again," escaped my lips. Two separate images filled my mind simultaneously.

The first was both a picture and a feeling of peddling [sic] my little sister on a bicycle like that on a summer day long ago in Indiana. I even felt the sweat on my face as we tackled the hill near our house. I smelled little Trish's hair, heard her laughter all over again. ...

The second image was of my having maybe killed the old man in front of the wall while taking out Rocket Man. That day had plagued me ever since. And, now, with the kids, it was like déjà vu, only ten times worse.[51]

Dave Grossman suggests that people find it harder to kill when they are closer to their victim, but the killings that anguished Matt Martin were executed from the same distance as the ones that gave him satisfaction. Physical distance is not the only or even the key variable that shaped his subjective experience of killing.

Like Elaine Scarry, Susan Sontag objects to the common assumption that we feel an automatic empathic discomfort at the bodily suffering of others, especially if it is witnessed close-up. "No 'we' should be taken for granted when the subject is looking at other people's pain," she wrote in her book *Regarding the Pain of Others*.[52] Sontag reminds us that soldiers sometimes take pleasure in killing and that white post-Reconstruction lynch mobs enjoyed the spectacle of black people being lynched, took their children to watch, and often kept souvenir pictures. In his much-cited essay "Why Men Love War," Vietnam veteran William Broyles also reports that there can be great joy in killing:

After one ambush my men brought back the body of a North Vietnam-ese soldier. I later found the dead man propped against some C-ration boxes; he had on sunglasses, and a *Playboy* magazine lay open in his lap; a cigarette dangled jauntily from his mouth, and on his head was perched a large and perfectly formed piece of shit. I pretended to be outraged, since desecrating bodies was frowned on as un-American and counterproductive. But it wasn't outrage I felt. I kept my officer's face on, but inside I was ... laughing. I laughed—I believe now—in part be-cause of some subconscious appreciation of this obscene linkage of sex and excrement and death; and in part because of the exultant realization that he—whoever he had been—was dead and I—special, unique I me—was alive.

Later in the essay, Broyles reports on a lieutenant colonel whose men fought off a night attack on his base: "That morning, as they were surveying what they had done and loading the dead NVA [North Vietnamese Army]—all naked and covered with grease and mud so they could penetrate the barbed wire—on mechanical mules like so much garbage, there was a look of beatific contentment on the colonel's face that I had not seen except in charismatic churches. It was the look of a person transported into ecstasy."[53]

As for drone crews, as we have seen in this chapter, they sometimes gather around the screen to exult in a distant death and sometimes feel undone by what they see on screen. The difference is not whether drone operators are near or close to their victims (deaths observed from the same distance on the same screen can result in jubilation or anguish) but whether the operators believe the target deserved to die. The deaths that haunt Martin are those of the two boys on the bicycle and the old man. "The Rocket Man had it coming. The old man did not."[54]

It is intrinsic to the phenomenon I have been calling remote narrativization that drone operators, acting from the space usually occupied by the gods, craft moral narratives about the people they watch from the heavens and the reasons they deserve to live or die. In the narratives of many drone operators, people who hide in the shadows and shoot at Americans on the ground deserve to die, not simply because they are on the wrong team, but, in keeping with the Manichean predilections of American political discourse, because they are "bad guys." American intervention in the Middle East is seen as a moral errand by the exceptional nation, not an exercise in colonial occupation, and this places insurgents on the wrong side of a moral divide. As framed by the dominant American ideology that people choose how to behave—who to be—and that these choices reflect their moral worth as individuals, insurgents are bad people whose punishment is righteous. They "had it coming." The drone operator Bill "Sweet" Tart, reflecting on his mindset as he prepared to kill someone from the heavens, said that "the seriousness of it is that I am going to do this and it will affect his family. But that individual is the one that brought it on himself."[55] If American leaders sometimes meld the laws of war and international criminal enforcement in their legal justifications for drone strikes (as is shown in chapter 5), the moral narratives of lowly drone operators who are smiting people from the heavens are often marked by the same fusion of the frames of war and righteous punishment.

The moral framing that is provided by processes of remote individualization is problematic, however. Judgments are made about people simply on the basis of their observed actions or their placement on an intelligence agency's target list for reasons that the drone operator often does not know. A judge handing down a punishment in a murder trial wants to know the mental state, family history, and possible extenuating circumstances of the defendant before pronouncing a death sentence. A drone operator who is limited to watching from afar and filling in the blanks knows none of this. As I argue in the next chapter, many Afghans and Iraqis become insurgents through a social logic that is driven by economic need, patronage relations, or clan loyalty, not because they are "Islamofascists" who oppose the ideals of the U.S Constitution (about which, in many cases, they know next to nothing). Drone operators are confronted with a local social logic that they often do not understand and which they then recode through the cultural logic of American moral individualism.

Remote Stress

It is commonly said that drone pilots have the same level of posttraumatic stress disorder (PTSD) as those who engage in combat, although the evidence is open to dispute. The diagnosis of PTSD is notoriously subjective, too few studies have been done, and those that have been done have been called into question. Nevertheless, a 2011 Pentagon study of 840 drone operators found that 46 to 48 percent experienced high levels of "operational stress," with 17 percent "clinically distressed," and 4 percent experiencing full-blown PTSD.[56] The classic symptoms of PTSD—nightmares, flashbacks, involuntary anger, hypervigilance, racing heart, and so on—are often seen as a sort of involuntary neurological response to situations of intense violence and danger. A person who develops PTSD might have been attacked in an ambush, survived an improvised explosive device detonation, or watched a friend killed by a sniper or bomb. A person with PTSD might seem persistently distracted, emotionally numb, but prone to outbreaks of anger; and certain sounds, smells, or sights can trigger sudden intense involuntary flashbacks.

Because drone operators do not have the direct, embodied experience of intense personal danger—bombs exploding and comrades shot down beside them—that marks the life of a soldier on the ground, some have expressed skepticism, sometimes even withering contempt, at the notion that drone operators might suffer from PTSD. The surgeon-general of the U.S. Air Force has stated that drone operators are at lower risk for PTSD than the general civilian population, and the air force tends to avoid the term *posttraumatic stress disorder*, referring instead to high levels of stress, which it attributes to the long shifts that drone operators work and to the difficulty drone operators experience in navigating the abrupt transition from killing on screen to engaging in domestic tasks, such as picking up their kids from a soccer game.[57]

This is certainly plausible. But there is evidence that drone operators can be psychologically scarred by helplessly watching on screen and listening on the radio as U.S. troops are killed by insurgents and, in some cases at least, as their own actions result in deaths. In his portrait of the drone operator Brandon Bryant (who eventually was diagnosed with PTSD), Matthew Power reports that, after his first kill, Bryant stopped his car while driving home and broke down sobbing. He was crying as he called his mother from the road to tell her that he had killed someone. He reported that a colleague drank an entire bottle of whiskey every time he killed someone and another "had nightmares after watching two headless bodies float down the Tigris." Another drone operator reacted to her first (and as it turned out, only) kill by risking court martial rather than fire again.[58] These may not be typical of all drone operators, but their reactions are real.

Journalist Chris Woods says that "psychologists are having to invent a new language to describe the damaging effects of this remote warfare on military personnel."[59] Traditional theories of PTSD present it as a neurological reaction to the experience of being attacked. However, some recent work has presented it as a "moral injury." In her ethnography of Iraq war veterans with PTSD, anthropologist Erin Finley reports that PTSD often seemed to correlate not so much with the absolute scale of raw violence but with its degree of senselessness or the degree to which veterans blamed themselves for a comrade's or civilian's death.[60] Throughout this chapter, I have described how drone operators construct narratives about the people they track and sometimes kill. These narratives often embody satisfying plotlines about enemy fighters who are justly dispatched to the next world. But they can be more troubling when old men or children die because they are in the wrong place at the wrong time, or operators are helpless witnesses to the slaughter of their own troops. Jonathan Shay, the psychiatrist who has led the move to reframe PTSD as a "moral injury," would argue that narratives are key to making sense of injury, suffering, and atrocity, and Finlay's work shows that narratives about unjust death are an important part of the dynamic of PTSD. Although drone operators may not experience the sounds and smells of combat that often trigger PTSD flashbacks, they carry with them intense visual images that can be hard to shake loose from their minds. If they do not fit the classic profile of a military veteran who develops PTSD, maybe this calls for a rethinking of PTSD—a syndrome that is poorly understood—rather than impugning the experiences of drone operators.

4 Casualties

It's not about the technology, it's about how the technology is used.
Drones aren't magically better at avoiding civilians than fighter jets.
—Sarah Holewinski[1]

The categories we take as rigid and unchanging, such as "terrorist," are
in fact remarkably fluid in the context of Afghan politics.
—Anand Gopal[2]

In February 2011, at a conference on drones and international
law organized by the New America Foundation, Tom Malinowski,
then the director of Human Rights Watch, surprised some in
the audience by making a strong defense of drone warfare.[3]
Malinowski argued that, compared to manned aircraft, drones
created the opportunity for more ethical and discriminate
attacks against targets on the ground because of their ability to
linger, often unseen, for hours as they track and evaluate potential targets. Pilots of manned planes flying at high speed and
working against time limits enforced by limited fuel might make
a rash targeting decision based largely on their own transitory
perceptions. Drones, however, can linger for hours as video feeds
from cameras are routed simultaneously to multiple decision

makers in the United States. These decision makers can consult military lawyers on the laws of war as they debate the pros and cons of attacking a particular target. As the drone circles high above the terrain, relaying high-resolution imagery to command centers away from the battlefield, decision makers can discuss their level of confidence that the potential target is an insurgent, even an individually known insurgent with a detailed case history, and that civilian casualties will be minimized if a missile is unleashed from the drone.

As Peter Singer has observed:

As recently as the 1999 Kosovo war, NATO pilots spotting for Serbian military targets on the ground had to fly over the suspected enemy position, then put their plane on autopilot while they wrote down the coordinates of the target on their lap with a grease pencil. They would then radio the coordinates back to base, where planners would try to figure out if there were too many civilians nearby. If not, the base would order an attack, usually made by another plane. That new plane, just arriving on the scene, would carry out the attack using the directions of the spotter plane, if they were still there, or the relayed coordinates. Each step was filled with potential for miscommunication and unintended errors. Plus, by the time a decision had been made, the situation on the ground might have changed—the military target might have moved or civilians might have entered the area.

Compare this with a UAV that can fly over the target and send precise GPS coordinates and live video back to the operators. Add in the possibility of using an AI [artificial intelligence] simulation to predict how many civilians might be killed, and it is easy to see how collateral damage can be greatly reduced by robotic precision.[4]

Similar arguments have been made by U.S. government officials about the ethical superiority of drone warfare. The strikes are often described by U.S. officials as "surgical" and "precise." The press has been told that President Obama holds regular "Terror Tuesday" meetings where individuals are added to the target list by name after the evidence against each individual has been carefully weighed. In February 2012, speaking in a YouTube forum, President Obama described drone strikes as "a targeted, focused effort at people who are on a list of active terrorists." He added, "I want to make sure that people understand: actually, drones have not caused a huge number of civilian casualties. ... For the most part they have been very precise precision strikes against Al Qaeda and their affiliates."[5] Secretary of State John Kerry has said that drones targeted only "confirmed terrorist targets at the highest level."[6] In April 2012, John Brennan, appearing before the Senate Intelligence Committee as the nominee for director of the Central Intelligence Agency, said "we only authorize a particular operation against a specific individual if we have a high degree of confidence that the individual being targeted is indeed the terrorist we are pursuing."[7] Earlier, in 2011, Brennan had claimed, remarkably, that U.S. drone strikes were so precise that they had not caused a single civilian death.[8]

These arguments are framed in terms of the Augustinian tradition of just war theory, which has been codified in the international laws of war. According to this tradition, killing in war is justifiable if attacks are guided by the principles of distinction, proportionality, and military necessity. According to these principles, attackers should take care to distinguish between military and civilian targets and to confine attacks to primarily military targets. It is permissible for these attacks to inflict civilian casualties as long as the primary intended target is military and the damage to civilians is not out of proportion to the military payoff. Targets should be selected not to terrorize entire populations indiscriminately or out of a desire for blood or revenge but to follow a logic of military necessity and proportionality. If two weapons would both destroy the same military target but one would cause fewer civilian casualties in doing so, there is an ethical and legal imperative to choose that weapon. And if an attack would cause far more civilian than military casualties, there is an onus of restraint on the attacker.[9]

Civilian Casualties

The claims of low civilian casualties made by U.S. leaders have intuitive force and plausibility in the context of a drone's technological capabilities—its ability to linger for hours over a target and relay live video footage to control rooms at headquarters, where lawyers and military personnel can discuss the deaths likely to be caused by an attack. Yet several research studies suggest that U.S. drone strikes leave significant numbers of dead civilians in their wake.

Given that drone strikes are aimed at hostile terrain that is largely inaccessible to U.S. military personnel and is dangerous even for independent observers, accurate assessments of the casualties caused by the strikes are difficult to undertake. A report by the Stanford Law School and the New York University School of Law lists some of the impediments that are encountered by independent observers who seek to catalog civilian casualties from drone strikes in the tribal areas of Pakistan. Army checkpoints obstruct travel in and out of this semiautonomous region, so that "trips that would normally take only a few hours can take days, or travelers may be turned back before they reach their destination." Residents are afraid of being killed by either government forces or the local Taliban for talking to outsiders in an environment where suspected spies can be summarily executed. And practices of purdah make it hard for men to know and tell outsiders the identities of women who are killed in drone strikes in neighboring compounds or even in their own extended family compounds.[10]

Even if it is possible to gain access to the site of a drone strike, there are further ambiguities. First, the victims' bodies may be so badly destroyed that it is hard for responders to know from the scattered and charred body parts how many people they represent. One eyewitness to a drone strike said that "their bodies were totally destroyed. ... We can't say that it is exactly four persons. It could be five or six as well because they were cut into pieces. We couldn't identify them."[11] Second, although the dead bodies of women and children, if they are recognizable as such, are usually taken as self-evident civilian casualties, it can be harder to determine whether a male corpse is that of an insurgent because insurgents do not wear military uniforms.[12] Some CIA analysts have complained that the White House casualty estimation protocol "counts all military-age males in a strike zone as combatants ... unless there is explicit intelligence posthumously proving them innocent." The journalist Glenn Greenwald quotes one CIA analyst as saying, "It bothers me when they say there were seven guys, so they must all be militants. They count the corpses and they're not really sure who they are."[13] On the other hand, insurgents have recruited teenagers as child soldiers and have been happy to play up civilian casualties, tweeting pictures of children's bodies as part of their own propaganda campaign.[14] In the end, "Journalists often find themselves in the position of having to choose between reporting 'official' casualty figures that they consider untrustworthy, or higher numbers from civilian sources that they may be unable to corroborate,"[15] and assessments of civilian casualties can pivot on the degree to which one is willing to trust local eyewitnesses and human rights activists versus anonymous government sources. To give a sense of the difficulty of precise estimation and the range of estimates in play, figure 4.1 shows estimates from three different sources of civilian casualties caused by drone strikes in Pakistan in 2011.

Despite differences in estimates, several studies undertaken by a range of researchers have challenged, to varying degrees, the low-casualty narrative of official sources. The London-based Bureau of Investigative Journalism, a nongovernmental organization that is critical of drone strikes, has invested considerable resources in tracking media accounts and descriptions on the ground of individual drone strikes. Its estimates of civilian casualties are the highest. It claims that U.S. drone strikes in Pakistan have killed between 2,438 and 3,942 people, with

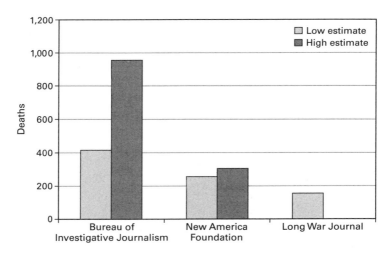

Figure 4.1
Counts of civilians who have been reported killed in Pakistan, 2011.
Source: Bureau of Investigative Journalism, UK.

civilians comprising between 416 and 959 of the total—in other words, between 10 and 40 percent. The Bureau estimates that at least 168, and possibly as many as 204, of the dead were children.[16]

The Washington-based New America Foundation, a think tank that is more supportive of drone strikes, has also sought to track and investigate each individual strike. It has a stricter protocol for confirming a casualty as civilian, putting less credence than the Bureau of Investigative Journalism in accounts by Pakistani journalists and human rights activists and more credence in U.S. and Pakistani government assessments. Its estimates of total deaths in Pakistan—between 2,227 and 3,612—are only a little lower than the Bureau's. But its estimates of civilian deaths are considerably lower, though hardly insubstantial, at 258 to 307 (7 to 14 percent).[17]

The lowest estimates are given by the *Long War Journal*, which is funded by the neoconservative Foundation for the Defense of Democracies and has strong ties to the U.S. military, especially the Marine Corps. It tends to rely heavily on conversations with unnamed U.S. intelligence officials as sources and to assume that drone victims are "militants" unless clearly proven otherwise.[18] But even the *Long War Journal* concedes 156 confirmed civilian deaths in Pakistan out of 2,903 total deaths, and the *Long War Journal*'s managing editor, Bill Rogio, has been quoted as saying that "the CIA's claim of zero civilian casualties in a year is absurd."[19]

A joint study by human rights lawyers from the law schools at Stanford and NYU investigated media coverage of drone strikes and aggregated data from 130 interviews with victims, witnesses, and humanitarian workers in Pakistan. The study does not give an independent estimate of civilian casualties but suggests a preference for the Bureau of Investigative Journalism's estimates – the highest of the three discussed above. The study says that U.S. government estimates of civilian casualties "are far lower than media reports, eyewitness accounts, and the U.S. government's own anonymous leaks suggest." It concludes:

In the United States the dominant narrative about the use of drones in Pakistan is of a surgically precise and effective tool that makes the U.S safer by enabling "targeted killing" of terrorists, with minimal downsides or collateral impacts.
This narrative is false.[20]

A UN special rapporteur, Ben Emmerson, also undertook a study of drone strikes in Pakistan, for which he gained access to a confidential study by government officials in Pakistan's federally administered tribal area (FATA). Emmerson estimated that between 400 and 600 civilians had been killed in drone strikes and singled out for particular condemnation one drone strike in 2006 that, he said, killed eighty children in a religious school.[21]

Perhaps most startling is a report by Larry Lewis of the Center for Naval Analyses, a center that is closely tied to the military. The classified study, based on U.S. intelligence assessments not shared with the public, was reported to have concluded that drone strikes in Afghanistan were ten times as likely as strikes by manned fighter jets to kill civilians. Because the study is classified, it is hard to assess its methods and reliability.[22]

Based on these studies, it seems clear that pronouncements by high U.S. government officials have misled the public about the number of civilians killed by U.S. drones, especially in Pakistan. Although we can argue about the exact number and proportion of civilians killed, it seems safe to say that they number considerably more than U.S. leaders have publicly conceded.

The picture that emerges from these studies by journalists, human rights activists, and independent military analysts is a far cry from the high-tech fairy tale in which lawyers and military operatives check and double-check drone video footage to ensure that no innocents are caught in the crosshairs before drone operators press the "pickle" button. The American people got dramatic confirmation of the gap between the ideal and the reality on April 22, 2015, when Barack Obama announced at a press conference that a recent CIA drone strike in Pakistan had destroyed a compound containing six people, not four as the CIA believed, and that the two unintended dead were an American and an Italian hostage whose presence was apparently undetected during hours of drone surveillance.[23]

What accounts for this disparity between ideal and reality? Given drones' capability to deliver precisely targeted strikes against individual adversaries and given the compelling plausibility of official accounts of targeting protocols, how have we ended up with so many dead women, children, and members of wedding parties? What has gone wrong?

I suggest that the answer to this question lies in a process of technical, organizational, and ethical slippage. This process of slippage has widened the discrepancy between idealized scripts about drone targeting and actual practices. It is unclear whether this discrepancy was apparent to U.S. government officials who misstated the accuracy and ethical defensibility of drone warfare. If the discrepancy between ideal and actual practices was known to them, then they deliberately misled the public. But It is possible that the power of formal scripts about targeting acted like a fog that prevented these officials from perceiving the space between the ideal and the real.

This process of slippage in the actual use of drones reminds us to be wary of what might be called "drone essentialism." In the social sciences, the word *essentialism* refers to the notion that all members of a category share an essence that makes them fundamentally alike. For example, those who essentialize gender assume that women are inherently different from men and share certain essential characteristics (such as a maternal instinct) by virtue of their gender. It is easy to be essentialist about weapons, too, by assuming that the consequences of their use are determined by their technical characteristics. Thus, it is often assumed that nonlethal weapons never kill people, that nuclear weapons are inherently genocidal, and so on. As Brian Rappert has shown, however, nonlethal weapons have killed people many times when used recklessly by the police.[24] And nuclear weapons can be used to destroy entire cities, to destroy a purely military target like a submarine, or as symbolic tokens in a game of deterrence. As for drones, they can be used as parsimonious instruments of violence, firing missiles only when there is near certainty that no one but a confirmed combatant will be killed, but they also can be deployed with looser targeting protocols. The same drone with the same video capability and the same missiles under the wing can kill more or fewer people and more or fewer innocent civilians depending on the orders that have come down through the chain of command, the training of drone operators, the pressures from commanders on the ground, and the organizational culture in which the drone team is embedded. A drone is a sociotechnical ensemble, not just a machine, and the same drone will be deployed to different effects in different cultural and organizational contexts. A simple example illustrating this argument is that the United States has used drones in regular war zones such as Iraq and Afghanistan and also in other countries, such as

Yemen, with which it is not at war. Until late 2015, when a British drone attacked a target in Syria, the United Kingdom, with a
different legal culture, confined its drone strikes to Iraq and
Afghanistan (and, later, Libya) arguing that the use of drones to
attack targets outside these established war zones would be a violation of international law.[25] The remainder of this chapter
shows that the targeting protocols adopted by the United States
have changed over time and that these changes have increased
civilian casualty rates substantially.

Targeting

Personality Strikes and Signature Strikes
Statements by President Barack Obama and other U.S. officials
have given the impression that drones attack only people whose
identity is known and whose names have been placed on a
target list of individuals who have been determined to be important insurgents or terrorists hostile to the United States. (According to some media accounts, the White House makes what are
nicknamed "baseball cards" with the pictures and condensed
biographical information of each targeted individual.)[26] These
are known in official parlance as "personality strikes" because
they are directed at individuals who have personalized profiles.
These high-value targets appear on separate but overlapping target lists that are maintained by U.S. military and intelligence
agencies and are vetted by interagency committees that bring
their nominations to the White House for approval at weekly
meetings.[27] According to investigative journalist Jeremy Scahill,
the process of building and approving a file on an individual to
be killed has typically taken just under sixty days, and authorization to attack someone on the "kill list" has to be renewed after

sixty days. Scahill claims that, in the days just before the bureau-cratic window closes, the temptation is higher to weaken stan-dards for confirming the identity of the target, or to attack despite the likelihood of civilian casualties.[28]

A 2010 investigation by Reuters found that drone strikes do not kill primarily these "high-value" people. Of the five hundred people that the CIA believed it had killed with drones over the previous two years, only fourteen were top-tier militant targets, and only twenty-five were mid- to high-level organizers. The report concluded that the CIA had killed twelve times as many low-level as mid- to high-level fighters.[29]

Furthermore, leaks to journalists have revealed that most drone strikes are not personality strikes, where the target's iden-tity is known, but instead are "signature strikes" (a term that was classified but is now widely known through media reports).[30] Signature strikes were proposed by CIA director Michael Hayden and approved by President George W. Bush in 2008 when increasing numbers of Taliban fighters were crossing the border from Pakistan to Afghanistan, unmolested by the Pakistani military; the adoption of this looser targeting protocol enabled a massive uptick in the rate of attack.[31] The targets of a signature strike (referred to by some intelligence officials as "crowd kill-ing")[32] do not appear on an official target list but have exhibited signature behaviors that are associated with insurgents. Although their identity is unknown, they can be killed based on behav-ioral profiling. And the kind of behavior that can help earn a sentence of death by Hellfire missile may be quite broad. CIA targeters reportedly believe that Pashtun men urinate standing up and that Arab men—who are more likely to be al Qaeda operatives—squat.[33] At particular risk are men of military age (anywhere from their late teens to their sixties).[34] (In 2010,

General Stanley McChrystal banned the term "military-aged males" because he feared that "it implied that every adult man was a combatant."[35])

Some people who work in the national security bureaucracy have protested that the protocols for signature strikes make it too easy to kill innocent people. According to the *New York Times*, which first reported the existence of signature strikes, State Department officials have claimed that "when the CIA sees 'three guys doing jumping jacks,' the agency thinks it is a terrorist training camp. ... men loading a truck with fertilizer could be bombmakers, but they might also be farmers."[36] The journalist Jonathan Landay worries about "a breadth of targeting that is complicated by the culture in the restive region of Pakistan where militants and ordinary tribesmen dress the same, and carrying a weapon is part of the centuries-old tradition of the Pashtun ethnic group."[37] The *New York Times* reports that counterterrorism officials defend signature strikes by saying that "this approach is one of simple logic: people in an area of known terrorist activity, or found with a top el Qaeda operative, are probably up to no good."[38]

Micah Zenko of the Council on Foreign Relations points out an irony that is inherent in the logic of signature strikes:

It is striking to compare Obama's deliberate and thoughtful commentary about the tragic killing of Trayvon Martin with the military tactic that will forever characterize his presidency: killing people with drones. The president posits that it is wrong to profile individuals based upon their appearance, associations, or statistical propensity to violence. By extension, he believes that, just because those characteristics may seem threatening to some, the use of lethal force cannot be justified as self-defense unless there are reasonable grounds to fear imminent bodily harm. But that very kind of profiling and a broad interpretation of what constitutes a threat are the foundational principles of U.S. "signature strikes"—the targeted killings of unidentified military-age males … if you apply Obama's logic concerning the Trayvon Martin tragedy, hanging around in the wrong neighborhood or with bad people should not make a person guilty.[39]

Double-Tap Strikes

One subtype of signature strike is a "double-tap strike." In a double-tap strike, a drone continues to circle over a site after an initial strike and then launches further attacks from the air against those who come to the aid of the victims of the initial strike on the assumption that these responders must be in league with the original targets. Following the same logic of guilt by association, double-tap strikes have also been made against the people who attend funerals of initial strike victims.[40] The double-tap tactic is common enough that one humanitarian organization has a policy of waiting for six hours before rendering assistance after a drone strike, and many people in tribal areas have stopped attending funerals. One drone strike survivor said, "When a drone strikes and people die, nobody comes near the bodies for half an hour because they fear another missile will strike."[41]

This form of attack is modeled on a tactic that is practiced by groups such as Hamas and has been condemned by human rights lawyers. Christof Heynes, a UN special rapporteur on extrajudicial killings and summary or arbitrary executions, has argued that the people who pull bodies out of the rubble may be innocent bystanders indulging a humane instinct to help the victims. Clive Stafford-Smith, a lawyer who heads the charity Reprieve, has said that double-tap strikes "are like attacking the Red Cross on the battlefield."[42] And a report issued by the law schools at Stanford and NYU says that "intentional strikes on first responders may constitute war crimes."[43]

Personality Strikes Revisited

It would be a mistake to conclude that only signature strikes and double-tap strikes result in civilian casualties. Deliberate strikes against more clearly identified targets—personality strikes—can also cause civilian deaths. The previous chapter discussed a drone operator, Matt Martin, who fired at an insurgent whom he nicknamed "Rocket Man" and then saw an old man totter into the frame seconds before the missile struck. In this case, the Augustinian tradition and the laws of war are clear. The drone operator and his superiors calculated that a wall around the yard would protect bystanders from the Hellfire missile's blast and took care to shield civilians from a legitimate strike against a clear combatant. Despite their best efforts, the old man wandered into the frame in the instant between the pressing of the button and the obliteration of the target. This was not their fault. The old man, victim to the worst imaginable timing, was collateral damage to a strike that was permissible under the laws of war.

In other, more ambiguous cases, U.S. drone operators and their commanders can see that civilians will be killed along with targeted insurgents, and they weigh the number of expected civilian casualties against the importance of the combatant or combatants they are trying to kill. Reportedly, if ten or fewer civilians are expected to die, the decision can be made relatively low in the chain of command. If it is anticipated that more than ten civilians will be killed, the decision is made higher up in the chain of command. In such calculations, according to an investigative team from *Der Spiegel*, "Bodyguards, drivers and male attendants were viewed as enemy combatants, whether or not they actually were. Only women, children and the elderly were treated as civilians. … If a Taliban fighter was repeatedly involved in deadly attacks, a 'weighing of interests' was performed. The military officials would then calculate how many human lives could be saved by the 'kill,' and how many civilians would potentially be killed in an airstrike."[44] The journalists say that the military personnel they spoke to see the latter guideline as a "cynical" way of stacking the books, although others might disagree.

There are also some identification errors in strikes on individual high-level militants who appear on the target list that is generated in the White House. The Human Rights organization Reprieve issued a scathing report that documents forty-one instances where high-level targets were named as killed in more than one drone strike. These men, said the report, "seemed to have achieved the impossible: to have 'died' in public reporting not just once, not just twice, but again and again. Reports indicate that each assassination target 'died' on average more than three times before their actual death." As Reprieve points out, this raises two questions: who is "the US killing in the first two strikes that miss their targets?," and "if the US intelligence is so poor that it is repeatedly missing its target, how can it know whether those killed are civilians?"[45]

The journalist Jonathan Landay describes one case of mistaken identity that resulted in the wrong person's death:

Information, according to one U.S. intelligence account, indicated that Badruddin Haqqani, the then-No. 2 leader of the Haqqani network, would be at a relative's funeral that day in North Waziristan. Watching the video feed from a drone high above the mourners, CIA operators in the United States identified a man they believed could be Badruddin Haqqani from the deference and numerous greetings he received. The man also supervised a private family viewing of the body.

Yet, despite a targeting process that the Administration says meets "the highest possible standards," it wasn't Badruddin Haqqani who died when one of the drone's missiles ripped apart the target's car after he left the funeral.

It was his younger brother Mohammad.

Friends later told reporters that Mohammad Haqqani was a religious student in his 20s uninvolved in terrorism.[46]

In this case, the identification of the victim was based on a plausible but misleading mix of circumstantial evidence, inference, and rumor.

When President Obama and other U.S. leaders speak of drone attacks on individual leaders of the insurgency who have been identified as engaged in planning or executing attacks on U.S. personnel, the impression given is that the identity of the person on the receiving end of the Hellfire missile is clearly known. It is worth pausing to ask how this might be the case. Drone cameras do not have high enough resolution to match the face of someone on the ground with a file photo (if one exists), and it is hard for the United States to get reliable agents close to insurgent leaders in places like Waziristan, Yemen, and Somalia where they might confirm a target's identity. In this situation, under its GILGAMESH program, the United States relies heavily on signals intelligence—specifically, cell phone data. "We Track 'Em, You Whack 'Em," is the informal motto of the team at the National Security Administration (NSA) that collects and analyzes this cell phone data.[47] A team of journalists from *Der Spiegel* describe the operational procedures involved:

Predator drones and Eurofighter jets equipped with sensors were constantly searching for the radio signals from known telephone numbers tied to the Taliban. The hunt began as soon as the mobile phones were switched on.

Britain's GCHQ and the US National Security Agency (NSA) maintained long lists of Afghan and Pakistani mobile phone numbers belonging to Taliban officials. A sophisticated mechanism was activated whenever a number was detected. If there was already a recording of the enemy combatant's voice in the archives, it was used for identification purposes. If the pattern matched, preparations for an operation could begin. The attacks were so devastating for the Taliban that they instructed their fighters to stop using mobile phones.

The document also reveals how vague the basis for deadly operations apparently was. In the voice recognition procedure, it was sufficient if a suspect identified himself by name once during the monitored conversation. Within the next 24 hours, this voice recognition was treated as "positive target identification" and, therefore, as legitimate grounds for an airstrike. This greatly increased the risk of civilian casualties.[48]

The journalists Glenn Greenwald and Jeremy Scahill describe one problem with using cell phones as proxies for individuals, based on a conversation that they had with a former drone operator:

One problem ... is that targets are increasingly aware of the NSA's reliance on geolocating, and have moved to thwart the tactic. Some have as many as 16 different SIM cards associated with their identity within the High Value Target system. Others, unaware that their mobile phone is being targeted, lend their phone, with the SIM card in it, to friends, children, spouses and family members. ...

"Once the bomb lands or a night raid happens, you know that phone is there," he says. "But we don't know who's behind it, who's holding it. It's of course assumed that the phone belongs to a human being who is nefarious and considered an 'unlawful enemy combatant.' This is where it gets very shady." ...

"People get hung up that there's a targeted list of people," he says. "It's really like we're targeting a cell phone. We're not going after people—we're going after their phones, in the hopes that the person on the other end of that missile is the bad guy."[49]

A whistleblower inside the drone targeting bureaucracy told the journalist Jeremy Scahill, "It's stunning the number of instances when selectors are misattributed to certain people. And it isn't until several months or years later that you all of a sudden realize that the entire time you thought you were going after this really hot target, you wind up realizing it was his mother's phone the whole time."[50]

The death of Zabet Amanullah, an Afghan political leader who had fought the Soviets and refused to join the Taliban, offers a dramatic example of the hazards of targeting cell phones as proxies for people. The U.S. military was seeking to kill Mohammed Amin, a senior Taliban figure whose name was high on the Joint Prioritized Effects List (JPEL), the list of people, approved by the White House, who are to be killed in signature strikes. But Zabet Amanullah's cell phone SIM card was incorrectly logged in U.S. databases as Mohammed Amin's SIM card, perhaps because of a malicious human informant or perhaps because of a data-entry error when Amanullah and Amin once spoke to one another on the phone. Although the owner of the phone that was being tracked by the U.S. military kept referring to himself on the phone as Zabet Amanullah and Zabet Amanullah was in the midst of a political campaign reported in Afghan media, the U.S. military persuaded themselves that Zabet Amanullah was a pseudonym for Mohammed Amin, and they killed him from the air during a campaign rally: "With eyes always on the telltale electronic signal, Amanullah's exuberant election rallies, the fifty-car convoy of well-wishers that escorted him to his home village, his pictures in the newspapers, his radio interviews, his daily phone calls to district police chiefs informing them of his movements—all passed the high-tech analysts by."[51]

We now know that there were disagreements within the North Atlantic Treaty Organization (NATO) about the criteria for determining that a cell phone signal confirmed the identity of an individual on the target list. German signals intelligence operatives, for example, insisted that the identity of the person who was holding the cell phone needed to be confirmed in real time rather than assumed, and they refused to pass on targeting information in the absence of such confirmation. Consequently, there were far fewer targeted killings in parts of Afghanistan for which German analysts had NATO responsibility.[52]

In addition to using cell phone signals to tag insurgent leaders, it is also widely rumored that the United States relies on informants on the ground, who are paid a bounty to identify militants and even mark their cars or homes with GPS tags that can guide drone strikes. According to former State Department official Lawrence Wilkerson, these bounties can go as high as $5,000, which is several years' income in Waziristan.[53] But we know from recent work by investigative journalists and others that such local informants can be notoriously untrustworthy and will sometimes fabricate allegations against unfortunate innocents or personal enemies to pocket the bounty or to settle personal feuds. As the Pakistani anthropologist Akbar Ahmed has observed, "Amid the confusion about the legitimacy of the targets, tribesmen with agnatic rivalry on their minds seemed to be playing their own devious games with the drones," sometimes "manipulating drone strikes to settle scores."[54]

Andrew Cockburn observes that "anyone who possessed one of these [GPS tag] devices held the power of life and death over anyone they chose. They could plant it in the home of an Al Qaeda terrorist or that of a neighbor with whom they were on bad terms."[55] In his remarkable book about the U.S. war in

Afghanistan, *No Good Men among the Living*, the investigative journalist Anand Gopal records several instances of Afghan tribal leaders who supported the U.S. invasion but were killed or sent to Guantanamo by U.S. troops on the basis of malign intelligence tips by local rivals. Here is his profile of an Afghan warlord called Gul Agha Sherzai, who made a business of parlaying false intelligence into a lavish lifestyle:

Eager to survive and prosper, he and his commanders followed the logic of the American presence to its obvious conclusion. They would create enemies where there were none, exploiting the perverse incentive mechanism that the Americans—without even realizing it—had put in place. Sherzai's enemies became America's enemies, his battles its battles. His personal feuds and jealousies were repackaged as "counterterrorism." ... Sherzai's network fed intelligence—which in the absence of an actual enemy was almost all false—to the Americans, and reaped the rewards: a business empire strung across the desert, garish villas abroad, and unfettered control of Southern Afghan politics. The Americans, in turn, carried out raids against a phantom enemy, happily fulfilling their mandate from Washington.[56]

Reportedly, 50 to 60 percent of those detained at the Abu Ghraib prison were innocent, and as many as 90 percent of those detained at Guantanamo were found never to have fought for al Qaeda.[57] How many victims of U.S. drone strikes were similarly misidentified?

An Expanded Target List

The Germans and the Americans also clashed over a U.S. attempt to expand the target list in Afghanistan. Bantz John Craddock, the U.S. general who served as NATO's Supreme Allied Commander for Europe, ordered strikes by ground troops and drones against drug dealers in Afghanistan on the grounds that the Taliban profited from the drug trade to the tune of at least $100 million a year. His order stated that being engaged in drug trafficking was sufficient to warrant attack and that individuals did not have to have demonstrated Taliban connections to be killed. Craddock's directive was classified, but it was leaked to the German press, and we now know that German officials, noting that the order had added thousands of Afghans to target lists, secretly protested that Craddock was violating both NATO rules of engagement and the international laws of war.[58]

Nor were Afghan drug dealers the only example of an expansion of the U.S. target list in a way that might surprise the American public. When American officials sought permission to operate in Pakistani airspace, as well as the cooperation of Pakistani intelligence officials in identifying people on the U.S. target lists, they agreed to use U.S. drones to kill Islamists that Pakistan saw as a threat to its own regime, although these men were not on the U.S. target list. In effect, the United States built its own set of baseball cards by trading cards with the Pakistanis.[59] The trading relationship was established by the very first U.S. drone strike in Pakistan, in June 2004, which killed Waziri tribal leader Nek Mohammad (along with two young boys and several others). The *New York Times* reported eight years later that "The C.I.A. had been monitoring the rise of Mr. Muhammad, but officials considered him to be more Pakistan's problem than America's. … [Mohammad] was not a top operative of Al Qaeda, but a Pakistani ally of the Taliban who led a tribal rebellion and was marked by Pakistan as an enemy of the state. In a secret deal, the C.I.A. had agreed to kill him in exchange for access to airspace it had long sought so it could use drones to hunt down its own enemies."[60]

Slippage

It is clear from investigations of civilian casualties on the ground and from details about operational targeting procedures (loose criteria for matching targets of personality strikes and the proliferation of signature strikes) that we should be skeptical of statements by President Obama and other U.S. leaders about parsimonious and strict targeting protocols in drone strikes. These statements are part of an official public script that is askew from the reality it purports to describe. This script presents an idealized representation of targeting practices from which actual practices have diverged as the United States has made side deals with Pakistan, lengthened its target lists, and increased the numbers of signature strikes. As the pace of drone strikes has intensified, there has been increasing slippage between rhetoric and reality. This slippage has been simultaneously operational and ethical. Why and how has this slippage taken place?

According to Mark Mazzetti, a *New York Times* intelligence beat reporter, part of the answer to this question lies in factional politics within the CIA, which is the agency primarily responsible for drone strikes in Pakistan.[61] In 2004, as the CIA was debating the merits of drone strikes in Pakistan, John Helgerson, the CIA inspector-general, issued a report that warned of the hazards associated with the CIA practice of detaining and torturing al Qaeda suspects or rendering them to countries such as Pakistan and Egypt, where they were tortured during interrogation. This had hitherto been a key tactic in identifying and neutralizing insurgent leaders. According to Mazzetti, "Mr. Helgerson raised questions about whether C.I.A. officers might face criminal prosecution for the interrogations carried out in the secret prisons, and he suggested that interrogation methods like waterboarding, sleep deprivation and exploitation of the phobias of prisoners—like confining them in a small box with live bugs—violated the United Nations Convention against Torture." Until then, according to Mazzetti, "the agency had been deeply ambivalent about drone warfare. The Predator had been considered a blunt and unsophisticated killing tool, and many at the C.I.A. were glad that the agency had gotten out of the assassination business long ago."[62] But the CIA leadership torch was being passed from the generation that had been scarred by the Senate's investigation of CIA assassination programs in the mid-1970s to a new generation that was more burdened by the scandals of torture and could contemplate the merits of targeted killing with fresh eyes: "The ground had shifted, and counterterrorism officials began to rethink the strategy for the secret war. Armed drones, and targeted killings in general, offered a new direction. Killing by remote control was the antithesis of the dirty, intimate work of interrogation. Before long the C.I.A. would go from being the long-term jailer of America's enemies to a military organization that erased them."[63] Ironically, within a few years some intelligence analysts would be lamenting that the turn to killing insurgent figures was denying them the opportunity to interrogate captured insurgents and harvest intelligence leads.

But the process of slippage was driven by more than CIA bureaucratic politics. It was also driven by the inner dynamics of counterinsurgency warfare itself. In his book *Dirty Wars* and its companion documentary film of the same title, Jeremy Scahill observes the paradox that the more people the United States kills with drones and special forces, the longer its target list becomes.[64] The United States was, in the words of Ryan Devereaux, "devoting tremendous resources to kill off a never-ending stream of nobodies."[65] This paradoxical reality— insurgency as a perpetual-motion mechanism—was the opposite of what the United States expected, which was insurgency as a form of motion that could be slowed to a crawl by the frictional force of counterinsurgency. As the U.S. embarked on its military campaign against insurgent networks in Iraq, Afghanistan, Pakistan, Yemen, and Somalia, it pictured these insurgencies as being led by a finite number of bad actors whose killing would decapitate and degrade their organizations. CIA director John Brennan spoke of drones' "surgical precision, the ability, with laser-like focus, to eliminate the cancerous tumor called an al-Qaida terrorist while limiting damage to the tissue around it."[66] At the same time, in keeping with the Manichean structure of thought that has informed much American public discourse about international conflict, especially after 9/11, these insurgents were imagined as unambiguous "bad guys" who deserved to be killed. The task was to separate the bad guys from the rest and kill them—to make a list and cross people off it until there was no more insurgency. This was often likened to "draining a swamp," a deeply misleading metaphor that is based on an assumption of finitude. It misses the actual dynamics of insurgency—a phenomenon that is more like a self-replicating immortal cell line than a swamp.

Social scientists and good journalists who have studied the social dynamics of insurgencies[67] tell us that the human boundaries of insurgent groups are often porous. Far from populations consisting of clear "good" and "bad" guys, some of their members often move in and out of insurgency depending on the economic opportunities available to them, shifting alliances in local tribal politics, ethnoreligious antagonisms, and the stimulus of atrocity, invasion, and subjugation. People may be part-time insurgents, accepting money from insurgent commanders to supplement other sources of income. In the context of tribal patronage relations, some young men may have no choice but to help local insurgents when a local leader reorients his loyalties in response to complex local jockeying for power. The journalist Steve Coll quotes a young man as saying that the local Pashtun communities "face irregular forces with long hair, beards, and their codes of conduct. It was very difficult to resist them. They imposed their own brand of Islam. If you did not cooperate, you were kidnapped, you were beheaded." Coll quotes another Pakistani who said that if the Taliban "comes to my *hujra* [guesthouse] and asks for shelter, you have no choice," even though it may be painting a bull's eye on one's home to allow the Taliban to stay there. This is partly because the Taliban are feared, and partly because of "tribal pressure to be hospitable. ... If you say no, you look like a coward and you lose face." As Coll observes, "In such a landscape, the binary categories recognized by international law—combatant or noncombatant—can seem inadequate. ... A young man of military age holding a gun outside a *hujra* might be a motivated Taliban volunteer, a reluctant conscript, or a victim of violent coercion."[68]

Men who had hitherto stayed on the sidelines may also take up arms in response to particular atrocities or incidents that reflect the cultural insensitivities of foreign occupiers or in a nationalist reflex against the indignity of invasion. "There is a perception of arrogance, there is a perception of helpless people being shot at like thunderbolts from the sky by an entity that is acting as though they have omniscience and omnipotence," said

General Stanley McCrystal, the former commander of U.S. forces in Afghanistan.

> You create a tremendous amount of resentment inside populations, not even the people that are themselves being targeted but around, because of the way it appears and feels. So I think that we need to be very, very cautious. What seems like a panacea to the messiness of war is not that at all. ... And wars are ultimately determined in the minds of populations.[69]

As counterinsurgency expert David Kilcullen has argued, if there is a way to prevail in counterinsurgency, it is by showing cultural sensitivity to occupied populations, creating widespread new economic opportunities, and refraining from the use of violence as much as possible. Instead, the United States has channeled economic aid to a corrupt few; burned down the opium crops on which many peasants rely for income, leaving them angry and destitute; used drone attacks to blow people apart from the skies; and trained troops to bash in the doors of family homes in the middle of the night, pointing guns at women and children, while screaming at them in English.[70] If one set out to create an insurgency, it is hard to imagine a set of policies better calculated to do so. And drone attacks are an integral part of the mix on which insurgency thrives. "Drones have replaced Guantanamo as the recruiting tool of choice," in the words of a *New York Times* article.[71]

"The U.S. has confused killing with winning," says Gregory Johnsen, an expert on Yemen. Interviewed on National Public Radio, Johnsen said of drone warfare, "The more men the U.S. killed, the stronger al Qaeda became. ... One of the fundamental truths of a war like this is that the side that kills the most civilians loses. Al Qaeda carried out a bloody assault on a hospital and, for days, people were up in arms. They were talking about what a horror, what a menace el Qaeda was. And then only a few days later the U.S. carried out a drone strike that seemed to be based on faulty intelligence. Instead of killing the target, the U.S. actually hit several cars that were in a wedding convoy and, just like that, all of the goodwill that the U.S. had garnered by el Qaeda making its mistake was lost. The difference in this is that el Qaeda apologized for the hospital attack. The U.S. never apologized for the wedding attack and, in fact, it continues to this day to say that it was a clean strike and that only terrorists were killed."[72] In a similar vein, the Yemeni writer Ibrahim Mothana has written that "the drone program is leading to the Talibanization of vast tribal areas and the radicalization of people who could otherwise be America's allies."[73] It is worth noting here that the suicide bomber who killed several Americans in the Khost compound in Afghanistan, the man who attempted to detonate a car bomb in Times Square, and the Afghan who plotted an attack on the New York subway system all claimed to be motivated by anger at U.S. drone strikes.[74]

American military planners who assumed that insurgent groups would be weakened and demoralized by the methodical elimination of their experienced local commanders and would give up or succumb to operational fracturing were often proved wrong. An internal Pentagon study by Rex Rivolo examined two hundred cases in 2007 to see whether, as the apostles of drone strikes would predict, fewer improvised explosive devices (IEDs) were detonated in the thirty days after local insurgent commanders were eliminated than in the thirty days before. He found, instead, on average a 40 percent increase in IED attacks within three kilometers of the former commanders' base of operations. He concluded that, in Andrew Cockburn's words, "new commanders were almost always eager to press the fight harder.

Often, they would be relatives of the dead man and hot for revenge. In addition, having just succeeded to the command, they would feel the need to prove themselves."[75] Instead of draining a swamp, U.S. attacks stirred a hornet's nest. U.S. counterinsurgency strategy has created a spiral within which drone strikes do not deplete the ranks of insurgents but instead help recruit new insurgents and intensify insurgent tactics. As the insurgency grows, military and intelligence officials who are convinced that drone strikes are the route to victory argue for relaxing targeting protocols and adding still more names to the death lists. Instead of draining the swamp, these drone strikes add fuel to the fire. But, seen as the solution rather than the problem, they become more and more favored as a military tactic. In the meantime, ethical discourses about discriminate, focused killing become increasingly out of joint with the practices that they mask and legitimate. They become threadbare alibis.

American decision makers congratulate themselves on using a more discriminate tool of violence and think that Pakistanis will surely see their moral superiority over the Islamist extremists who behead civilians or the Pakistani soldiers that often bomb indiscriminately. But this is often not the way local tribespeople see drone attacks, no matter how much many of them dislike the Taliban. For many living in Pakistan's tribal areas, in the crosshairs of the drones, it is precisely the deliberate quality of drone attacks that is resented. As one Pakistani survivor of a drone strike put it, "Drone strikes are not like other battles where innocent people are accidentally killed. Drone strikes target people before they kill them. The United States decides to kill someone, a person they only know from a video. A person who is not given a chance to say—I am not a terrorist.

The US chose to kill my mother."[76] And, as Steve Coll points out, "the relative precision of the aircraft that assailed them wasn't the point. On the ground ... drone war doesn't feel much different from other forms of air war, in that many civilians are displaced and frightened, and suffer loss of life and property."[77] To these tribespeople, the problem is not one side or the other so much as the war in its entirety, and insofar as U.S. drone warfare intensified that war, it was an integral part of the problem.

This process of slippage was not irreversible. U.S. drone strikes in Pakistan peaked in 2010 and then fell dramatically (figures 4.2 and 4.3). After 2011, civilian casualties in drone strikes in Pakistan decreased significantly both because the number of drone strikes was reduced and because the methodology of the strikes shifted in subtle but important ways. The United States reduced the number of drone strikes after three events in 2011 put enormous strain on U.S. relations with Pakistan. The first was an incident where Raymond Davis, the undercover CIA station chief in Lahore, shot and killed two men, reportedly undercover Pakistani agents, in traffic. The second was a drone strike in March, in which the U.S. killed dozens of tribal elders who had gathered for a dispute-resolution session on the assumption that such a large number of military-aged males gathered together were surely up to no good. "These people weren't gathered for a bake sale," one tin-eared U.S. official scoffed afterward. The third was the special forces raid, executed without the permission of Pakistan's government, that killed Osama bin Laden inside Pakistan. When Cameron Munter, the U.S. ambassador to Pakistan, warned that relations with Pakistan were deeply frayed, the Obama administration slowed the drone strikes.[78]

Meanwhile, the character of the strikes was changing. As the United States shifted from the Predator drone to the more

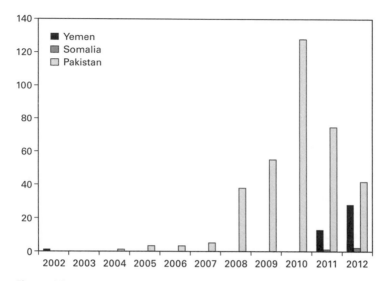

Figure 4.2
Number of U.S. drone strikes by year in Pakistan, Yemen, and Somalia, 2002 to 2012.
Source: Adapted from https://www.thebureauinvestigates.com/wp-content/uploads/2012/11/Every-confirmed-US-drone-strike-in-Pakistan-Yemen-and-Somalia-recorded-by-the-Bureau-2002-20121.jpg.

versatile Reaper, it could use smaller, more discriminate weapons. After 2010, in order to reduce civilian casualties, the CIA began to target cars suspected of carrying insurgents more often than houses.[79] In May 2013, under increasing pressure from human rights groups and some European countries, the White House announced new, tougher rules for drone strikes, which would now be allowed only if there was "near certainty that no civilians would be killed or injured, the highest standard we can set."[80] This was taken to imply restricting, if not completely discontinuing, the practice of signature strikes. After the strike that

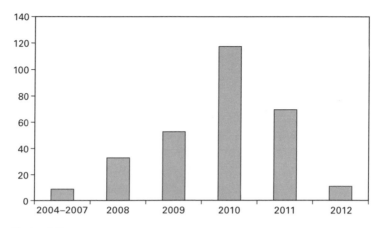

Figure 4.3
Drone strikes in Pakistan's tribal areas, 2004 to 2012.
Source: Adapted from http://www.cnn.com/2012/03/27/opinion/bergen
-drone-decline.

killed two Western hostages in January 2015, the head
of the CIA's Counterterrorism Center—a powerful advocate
for drones codenamed "Roger" who had pushed the number of
strikes sharply upward—was removed.[81]

Conclusion

Phrases such as "the ethics of drone warfare" suggest that drone
warfare has an essence, a stable form that we can determine to be
either ethical or unethical. As the literature in science and technology studies shows, however, technologies never have a single
frozen or ideal form: they can be mobilized in different ways by
different actors. As with nonlethal weapons, which can be lethal
in the hands of certain police officers, so with drones. They hold

out the promise of a more discriminate form of warfare that will kill fewer civilians, but this is a possible, not an assured, outcome of the use of drones. They also hold out the possibility of excessive killing with impunity. When insurgencies are not quickly contained with more discriminate forms of drone warfare, pressure builds for less discriminate targeting protocols, but the idealized image of immaculate targeting obscures the ethical shadings and shortcuts with which drone warfare becomes increasingly entangled.

5 Arsenal of Democracy?

Compared to Vietnam or World War II, wars that involved or convulsed all of American society, our "forever war" seems like an anomaly. But it would have been quite familiar to a nineteenth-century Briton: For these are the border wars of empire, which can never be won because no empire is ever free from threats.

—Adam Kirsch[1]

The Official Story

American government officials have not fully informed the American people about the U.S. conduct of drone warfare. They prefer the American people to be largely unaware of the drones patrolling faraway frontiers in their name. The locations from which drones fly, the targeting protocols used by drone crews, and the legal reasoning used to legitimate drone strikes—all have been closely held by the government, and until recently, the very term *signature strike* was classified. Not until three years after Anwar al-Awlaki, an American citizen, was deliberately killed in a drone strike in Yemen did the Obama administration publicly concede that it had ordered his death. The American Civil Liberties Union's requested information about the selection of drone targets under the Freedom of Information Act but was rebuffed. A *New York Times* reporter complained that "over the Obama presidency, it has become harder for journalists to obtain information from the government on the results of particular strikes. And Mr. Obama's Justice Department has fought in court for years to keep secret the legal opinions justifying strikes."[2] Even Harold Koh, the Yale Law School dean who became a principal architect of the government's legal justification for "targeted killings," complained in a speech shortly after he stepped down as legal adviser to the State Department that the Obama administration's "persistent and counterproductive lack of transparency" was responsible for "a growing perception that the [drone] program is not lawful and necessary, but illegal, unnecessary, and out of control."[3]

About a decade after the first drone strike, senior U.S. government officials began what seems to have been a coordinated strategy of making speeches at selected venues to explain and defend the U.S. government's use of drones. I begin this chapter with an exposition and analysis of four principal speeches. Often contain strikingly similar language, they present a composite portrait of the official case that the United States uses drones in accordance with the principles of domestic and international law to defend American democracy against the threat of terrorism. The speeches are President Obama's May 2013 speech, broadcast live on national television, at National Defense University;[4] Attorney-General Eric Holder's March 2012 remarks at Northwestern University Law School;[5] CIA Director John Brennan's May 2012 speech at the Woodrow Wilson Center;[6] and Harold Koh's March 2010 remarks to the American Society of International Law.[7] (After Koh left government for a position at the New York University

School of Law, over four hundred people, many of them NYU law students and teachers, signed a petition asking the university to cancel Koh's academic appointment because of his role in "crafting and defending what objectively amounts to an illegal and inhumane program of extrajudicial assassinations and potential war crimes.")[8]

All four speeches take as their starting point that the United States is using drones to defend freedom and democracy against its enemies in the war on terror. Eric Holder quotes President John F. Kennedy: "in the long history of the world, only a few generations have been granted the role of defending freedom in its hour of maximum danger," adding "it is clear that, once again, we have reached an 'hour of danger.'" Barack Obama's speech goes further than the others in acknowledging that an indefinite war on terror poses a threat to American democracy and in making the case that the terrorist attacks of 9/11 created a state of emergency in the international system that calls for exceptional measures until the threat represented by al Qaeda and other similar organizations is brought under control. President Obama cites James Madison's admonition that "no nation can preserve its freedom in the midst of perpetual war" but also says that "our commitment to constitutional principles has weathered every war, and every war has come to an end." Speaking of the attacks of 9/11, he says, "This was a different kind of war. No armies came to our shores, and our military was not the principal target. Instead, a group of terrorists came to kill as many civilians as they could." In his speech, the president presents drones as a key technology in the struggle against a new kind of adversary who does not fight on the regular battlefield but hides in remote places beyond the normal reach of the military: "They take refuge in remote tribal regions. They hide in caves and walled compounds. They train in empty deserts and rugged mountains."

Acknowledging that drones pose a challenge to democratic oversight, President Obama says that "the very precision of drone strikes and the necessary secrecy often involved in such actions can end up shielding our government from the scrutiny that a troop deployment invites." He says that "the technology to strike half a world away also demands the discipline to constrain that power," adding that "clear guidelines, oversight and accountability" have now been "codified." In a similar vein, Eric Holder says that "just as surely as we are a nation at war, we also are a nation of laws and values. Even when under attack, our actions must always be grounded on the bedrock of the Constitution—and must always be consistent with statutes, court precedent, the rule of law and our founding ideals."

Both Holder and Obama stress that congressional committees are briefed on drone strikes, ensuring some measure of congressional oversight. And all four speeches argue that, in terms of U.S. constitutional law, the authority to use drones to kill people in locations such as Afghanistan, Pakistan, Somalia, and Yemen derives from the Authorization for Use of Military Force (AUMF) against Terrorists of 2001 (enacted on September 14, 2001). Preempting objections that the United States is formally at war only in Afghanistan, and therefore should not kill people in other countries, Eric Holder says that the battlefield is wherever the enemy is: "Our legal authority is not limited to the battlefields of Afghanistan. Indeed, neither Congress nor our federal courts has limited the geographic scope of our ability to use force to the current conflict in Afghanistan. We are at war with a stateless enemy, prone to shifting operations from country to country."

The four speeches make the point that international law, particularly the UN charter, allows nations to use force to protect themselves and that they can act without international agreement against "imminent" threats or in self-defense. Holder says that "this does not mean that we can use military force whenever or wherever we want. International legal principles, including respect for another nation's sovereignty, constrain our ability to act unilaterally." Hinting that countries such as Pakistan and Yemen may have quietly given permission for drone strikes on their territory, he says that "the use of force in foreign territory would be consistent with these international legal principles if conducted, for example, with the consent of the nation involved." But he then justifies a much broader use of drones, even in the territory of countries that have not consented, by saying that drones also could be used "after a determination that the nation is unable or unwilling to deal effectively with a threat to the United States." Implicitly invoking the U.S. role as a global policeman, President Obama summarizes the argument in his own speech: "we act against terrorists who pose a continuing and imminent threat to the American people, and when there are no other governments capable of effectively addressing the threat." Making an implicit comparison to his predecessor in the Oval Office, he presents drone strikes as a minimalist, restrained, judicious way of dealing with threats—a scalpel rather than a hammer—compared to the blunt force often used by the George W. Bush administration: "Conventional airpower or missiles are far less precise than drones, and are likely to cause more civilian casualties and more local outrage. And invasions of these territories lead us to be viewed as occupying armies, unleash a torrent of unintended consequences, are difficult to contain, result in large numbers of civilian casualties and ultimately empower those who thrive on violent conflict."

The international laws of war require that targets should be primarily military and that any civilian casualties should be secondary and in proportion to the military value of a strike. In regard to this issue, Obama said that for a drone strike to go ahead, "there must be near certainty that no civilians will be killed or injured—the highest standard we can set." In CIA director John Brennan's words, "By targeting an individual terrorist or small numbers of terrorists with ordnance that can be adapted to avoid harming others in the immediate vicinity, it is hard to imagine a tool that can better minimize the risk to civilians than remotely piloted aircraft." Expressing some frustration with public attacks on drones, Harold Koh says that "it makes as little sense to attack drone technology as it does to attack the technology of such new weapons as spears, catapults, or guided missiles in their time. Cutting-edge technologies are often deployed for military purposes; whether or not that is lawful depends on whether they are deployed consistently with the laws of war, *jus ad bellum* and *jus in bello*. Because drone technology is highly precise, if properly controlled, it could be more lawful and more consistent with human rights and humanitarian law than the alternatives."[9]

John Brennan's speech is the only one that provides details about how targets are selected. Speaking as if all strikes were "personality strikes" on people whose names appear on lists of named targets (in a way that was challenged from the floor by a journalist for the Japanese newspaper *Asahi Shimbun*),[10] he emphasizes the detailed sifting of intelligence information about potential targets. In testimony that is hard to reconcile with the descriptions of actual drone strikes against unknown targets of opportunity (described earlier in this book), he concludes that "we only authorize a particular operation against a specific individual if we have a high degree of confidence that the individual being targeted is indeed the terrorist we are pursuing." Contesting allegations that "the Obama Administration somehow prefers killing al-Qaida members rather than capturing them," he also said that "our unqualified preference is to only undertake lethal force when we believe that capturing the person is not feasible."[11]

Koh, Holder, and Brennan all take issue with calling "targeted killings" "assassinations." (President Gerald Ford issued an Executive Order against assassinations in 1976, following revelations in the media and in congressional hearings that the CIA had targeted foreign leaders, such as Fidel Castro of Cuba and Patrice Lumumba of the Congo, for assassination.) "Some have called [drone] operations 'assassinations,'" says Holder: "They are not, and the use of that loaded term is misplaced. Assassinations are unlawful killings." All three speeches point out that the United States deliberately shot down a plane carrying Admiral Isoroku Yamamoto, the architect of Pearl Harbor, during World War II and that, in Koh's words, "this was a lawful operation then, and would be if conducted today."

In September 2011, a drone crew deliberately killed Anwar al-Awlaki, a U.S. citizen who had thrown in his lot with al Qaeda in Yemen. An imam whose lectures and writings encouraged jihad against the U.S., he was a highly effective recruiter and propagandist for al Qaeda, and U.S. officials claimed he had also become an operational planner of terrorist attacks as well. In their speeches, both Obama and Holder defend the deliberate killing of an American without a trial. Invoking the urgency of an imminent threat, which gives nations considerable latitude to defend themselves under international law, Obama says that "when a U.S. citizen goes abroad to wage war against America and is actively plotting to kill U.S. citizens, and when neither the United States nor our partners are in a position to capture him[12] before he carries out a plot, his citizenship should no more serve as a shield than a sniper shooting down on an innocent crowd should be protected from a SWAT team." Responding to claims that, as a U.S. citizen, al-Awlaki should not have been targeted without a court order, Holder said, "The Constitution guarantees due process, not judicial process."

In sum, the government's case is that drones offer a unique capability for protecting the United States and the liberal international order against terrorists who gather and plot in remote, lawless parts of the world; that the use of drones is consistent with U.S. domestic law and with the international laws of war; that drones are less likely to kill civilians than other military technologies; and that each drone strike is carefully reviewed to ensure that it is legally and ethically defensible. Drones are portrayed as discriminate weapons used sparingly and judiciously against rogue outlaws in the international system.

The Critics

Mainstream institutions in the United States have been largely silent about the drone wars that have taken shape on the frontiers of the American empire. America's churches have, on the whole, ignored the issue of targeted killing. Congress, deferring to the executive branch on this issue, has held almost no hearings and passed no laws on drone warfare. The courts have shown little interest, and one court turned away an appeal by Anwar al-Awlaki's father, before his son's killing, to force the

U.S. government to justify his inclusion on a death list.[13] In the words of John Kaag and Sarah Kreps, "The legislative and judicial branches have largely stepped aside rather than introduce roadblocks into the policy. Taken together, the checks and balances theoretically associated with democratic institutions have been inoperative."[14] Multiple opinion polls show that, although the general public in almost every other country in the world disapproves of drone strikes, over 60 percent of Americans support the use of drones for targeted killing outside the United States.[15]

Insofar as there have been critiques of drone warfare, they have come from international lawyers, nongovernmental organizations, some journalists, and peace and human rights activists. Three United Nations special rapporteurs have issued reports that express reservations about the legality of drone warfare as it is currently practiced.[16] In 2012, international human rights lawyers at Stanford and New York University law schools issued a joint report, *Living under Drones: Death, Injury, and Trauma to Civilians from U.S. Drone Practices in Pakistan*, that was based partly on interviews in the tribal areas of Pakistan. It criticized "the negative effects US policies are having on civilians living under drones" and called for "significant rethinking of current US targeted killing and drone strike policies."[17] Amnesty International also issued a report, which said that it was "seriously concerned" that drone strikes "have resulted in unlawful killings that may constitute extrajudicial executions or war crimes."[18] The Bureau of Investigative Journalism and Reprieve, both based in the United Kingdom, have partnered with a network of Pakistani lawyers and activists to produce a steady stream of claims that U.S. drone strikes kill large numbers of civilians.[19] And numerous books by activist journalists indict drone warfare as a way of killing people outside the law.[20] At the same time, main-stream journalists who write for newspapers such as the *Washington Post*, the *New York Times*, the *Los Angeles Times*, and *The Guardian* have uncovered facts that are often at odds with official representations of drone warfare.

There have also been public protests against drones. The largest, in Pakistan, were led by the politician and former celebrity cricket player Imran Khan who organized protests that mobilized thousands of people.[21] Small, sporadic protests have been held at U.S. Air Force bases and elsewhere in the United States. Although tiny compared to the mass mobilizations against the Vietnam War in the 1960s, against the escalation of the nuclear arms race in the 1980s, and against the Iraq War in 2003, they allowed for the performance of public moral witness. Many of these protests have been organized by the feminist group Code Pink, whose founder, Medea Benjamin, was able to interrupt (until she was forcibly removed from the room) President Obama's speech at National Defense University and CIA Director John Brennan's speech at the Wilson Center (figure 5.1).[22]

I fortuitously stumbled upon one of these protests when taking my children to the Smithsonian Air and Space museum in Washington one Saturday morning. We found a line of tourists waiting to get in, looking with bemusement at about thirty protestors, some dressed in bright pink, hoisting crude cardboard drones, unfurling banners, and giving impromptu speeches with the aid of a microphone and a portable loudspeaker (figure 5.2).

The speakers took turns at the microphone to lay out their claims: companies such as General Atomics and Raytheon are "making a killing out of killing"; drones kill more civilians than insurgents; children in the tribal areas of Pakistan have high

Figure 5.1
Medea Benjamin being removed from the room where President Obama is speaking.
Source: From https://tbmwomenintheworld.files.wordpress.com/2015/06/gettyimages-169345815.jpg.

rates of posttraumatic stress disorder from the buzzing of drones and the constant possibility of being dismembered without warning; and as a constitutional lawyer, President Obama should know that he cannot use drones to kill people in countries with which the United States is not at war and to kill American citizens without trial. The speeches were punctuated by periodic chants such as "When drones fly, children die" and "Hey, hey, ho, ho: killer drones have got to go."

"We're here in front of the Air and Space Museum," said one speaker, "because there's this giant drone exhibit inside. And we want you all to know the truth about the killer drones. Many of us here have been to Pakistan and met with drone strike

Figure 5.2
Protest at the National Air and Space Museum of the Smithsonian Institution. Photo by Hugh Gusterson, October 4, 2014.

survivors, and their tales are absolutely heartbreaking." Another speaker called the museum exhibit "war propaganda" and said that "The drone display inside that building doesn't educate. It shows you what a drone looks like, but it doesn't tell you what drones do."

Much of the critique of drones has turned on issues of international and domestic U.S. law. The critics fear that established laws of war and the aspirational norms that they enshrine are being eroded by drone warfare and that Daniel Reisner, former head of the Israeli Defense Forces legal department, was only too right when he said that "if you do something for long enough, the world will accept it. ... International law progresses through violations."[23] Philip Alston, a UN special rapporteurs, laments "a highly problematic blurring and expansion of the boundaries of the applicable legal frameworks—human rights law, the laws of war, and the law applicable to the use of inter-state force. Even where the laws of war are clearly applicable, there has been a tendency to expand who may permissibly be targeted and under what conditions."[24] James Joyner, writing in the *National Interest*, puts it more bluntly: "For centuries, civilized societies have understood that even wars must be fought according to rules, which have developed over time in response to changing realities. Rules are even more important in endless, murky wars such as the fight against Islamist terror groups. Currently, we're letting whomever is in the Oval Office pick and choose from among the existing rules, applying and redefining them based on his own judgment and that of his advisors."[25] Critics of drone warfare particularly worry that, by respatializing war, the U.S. use of drones effectively dissolves that part of international law, codified in the Geneva Conventions, that draws a reasonably clear line between the battlefield (where people can be killed) and civilian spaces (where they cannot) and between combatants and civilians. They also argue that drones make it too easy for presidents to act unilaterally and unaccountably.

The critics make the case that, outside the established battlefields of Iraq and Afghanistan at least, the president's use of drones has violated the U.S. War Powers Act of 1973 and the UN charter. The War Powers Act allows presidents to take military action in emergencies but reserves ultimate authority over the declaration of war to Congress. As for international law, according to the UN charter, nations can legitimately use force on the territory of other nation-states only if they are authorized to do so by the United Nations; if they have been attacked; if they are acting to preempt an imminent attack on themselves; or if they have the consent of the other state to do so. In fighting, they are supposed to discriminate between civilians and legitimate targets (those engaged in "continuous combat"), and "if there is doubt whether a person is a civilian, the person must be considered a civilian."[26] They can still take preemptive action against threats that are too intermittent and discontinuous to rise to the formal status of war. But operating within the frame of police action against a criminal rather than military action against an enemy, they have an obligation to do their best to capture, not kill. Obviously, a drone is not optimal for law enforcement purposes because it is designed for killing, not capturing.

Let's review these issues one by one.

War and the Law in the United States   According to the War Powers Act of 1973, the president must notify Congress within forty-eight hours after sending U.S. forces into combat and obtain congressional approval for combat operations that continue beyond sixty days. The Obama administration, trying to have it both ways, has argued that it has not needed permission from Congress to use drones outside the war zones in Iraq and Afghanistan for two different reasons: first, they are not wars; and, second, even if they are wars, they were already authorized by Congress in 2001. The administration has argued that the War Powers Act did not apply to the deployment of air power against Libya because the United States had no ground troops at risk of dying in those countries and therefore was not engaging in war. "U.S. operations do not involve sustained fighting or active exchanges of fire with hostile forces, nor do they involve U.S. ground troops," the White House stated, while Harold Koh averred that "the limited nature of this particular mission is not the kind of 'hostilities' envisioned by the War Powers

Resolution."[27] This position was contested by a number of lawyers, the most notable of whom was Jack Goldsmith, former head of the Office of Legal Counsel in the George W. Bush administration's Justice Department. Criticizing Obama's "astonishing legacy of expanding presidential war powers," he complained that it was bizarre to say that "thousands of air strikes that killed thousands of people and effected regime change" did not add up to war. He added that "the administration's theory implies that the president can wage war with drones and all manner of offshore missiles without having to bother with the War Powers Resolution's time limits."[28] Although phrased in a legal register, the issues raised here intersect with questions raised earlier in this book about whether drone pilots should be seen as having served in combat (given that they are unlikely to be killed), whether they deserve medals, and whether they can plausibly be said to suffer from posttraumatic stress disorder. In the law as in so many other respects the liminal quality of drone operations destabilizes existing categories. We will return at the end of the chapter to the question at issue between Harold Koh and Jack Goldsmith: are drones engaged in something readily recognizable as war?

Many lawyers have been no more impressed by the Obama administration's second argument, which is that the congressional Authorization for Use of Military Force (AUMF) of 2001 allowed drone attacks in Libya, Syria, Pakistan, and Yemen a decade later. In language that begs close and careful reading, the 2001 AUMF authorizes the president "To use all necessary and appropriate force against those nations, organizations, or persons he determines planned, authorized, committed, or aided the terrorist attacks that occurred on September 11, 2001, or harbored such organizations or persons, in order to prevent any future acts of international terrorism against the United States by such nations, organizations or persons."[29] CIA director John Brennan points out that "there is nothing in the AUMF that restricts the use of military force against al-Qaeda to Afghanistan."[30] On the other hand, the AUMF language clearly targets those responsible for 9/11. Although the AUMF may not have limited the use of force to Afghanistan, it is hard to imagine that, for example, Congress thought that it was authorizing an attack on the regime of Muammar Ghadaffi, an avowed foe of al Qaeda,[31] in Libya a decade later. In reference to Obama's invocation of AUMF to legitimate attacks on the new insurgent group the Islamic State in Iraq and Syria (ISIS) in 2014, Robert Chesney, a law professor at the University of Texas, said that "on its face this is an implausible argument because the 2001 AUMF requires a nexus to al Qaeda or associated forces of al Qaeda fighting the United States. ... Since ISIS broke up with al Qaeda it's hard to make that argument." Similarly, Benjamin Wittes of the Brookings Institution said that "surely associated forces doesn't mean forces that are actively hostile and have publicly broken with and been repudiated by al Qaeda. Whatever 'associated' means, I don't think it means that."[32]

A number of legal commentators in the United States were particularly upset by the targeted killing of a U.S. citizen, Anwar al-Awlaki, on September 30, 2011. Although there can be little doubt that al-Awlaki had made al Qaeda's cause his own, his government—acting as prosecutor, judge, jury, and executioner—killed him without arguing his guilt in court and without proving the case that his acts went beyond anti-American speech that might, despite its odious content, be protected by the First Amendment. Al-Awlaki preached vitriolic sermons against the United States, helped produce the al Qaeda magazine *Inspire*, and was associated with a number of people who attempted or carried out attacks on American targets—two of the 9/11 hijackers, the U.S. Army psychiatrist (Nidal Malik Hasan) who shot thirteen people at the Fort Hood military base, and the Nigerian "underwear bomber" (Umar Farouk Abdulmutallab) who attempted to destroy a U.S. jetliner on Christmas day 2009. The United States also claimed that in addition to his incendiary speech acts and unsavory associations, al-Awlaki had become an "operational commander" of al Qaeda, although it has not made public its evidence for this claim.

The critics insist that it was illegal for al-Awlaki's government to deliberately kill him without judicial process. Most commentators agree that American law allows only two circumstances in which the government can kill a U.S. citizen without a trial: "In U.S. cities police may shoot and kill suspects who present a risk to officers or bystanders. On the battlefield, U.S. troops do not need to examine the passports of those who are firing at them before firing back."[33] David Dow, a law professor at the University of Houston, notes that in both instances "there is a legal principle at work: immediacy. When a hostile actor presents an 'immediate' threat, the Constitution does not disable authorities from responding to that threat with lethal force."[34] But as the Georgetown law professor David Cole points out, it is hard to take seriously the Obama administration's insistence that al-Awlaki posed an immediate threat because there was a fifteen-month lag between the legal memo that authorized his assassination and the drone strike that killed him: "Can al-Awlaki really have posed an 'imminent' threat for the entire fifteen months between the time the memo was written and his killing, and if so, what does that tell us about the administration's conception of 'imminence'?"[35] As Dow says, "If al-Awlaki presented an immediate threat, then 'immediate' means anything, and therefore it means nothing at all."[36] Like many other legal commentators, Dow and Cole argue that the Obama administration violated fundamental precepts of the Constitution by issuing a death warrant without a judicial process that would allow the accused to know the evidence against him and defend himself. "Until last September," says Dow, "the rule of law in America seemed to mean that, at the very least, U.S. citizens could not be assassinated on the sole basis that the President has unilaterally concluded that killing them would make us safer."[37] "How can we be free if our government has the power to kill us in secret?" asks David Cole.[38] "If you believe the President should have the power to order people, including US citizens, executed with no due process and not even any checks or transparency, what power do you believe he shouldn't have?" asks the lawyer and author Glenn Greenwald.[39]

War and the Law in the International System   Critics have also challenged the legal basis for drone strikes in international law. Under the UN charter, these strikes could be justified if conducted with the consent of the government whose territory was struck or in response to a direct attack or an imminent threat of attack. The U.S. government has invoked both justifications. With reference to Pakistan, U.S. officials have let it be known to journalists that Pakistan's government has condemned U.S. drone strikes in public but has approved and actively enabled them in private. There is good reason to believe this claim.[40] But whatever deals their government might have struck behind closed doors, opinion polls show that large majorities of Pakistanis oppose the drone strikes,[41] and in 2012 both houses of Parliament in Pakistan voted to demand an immediate end to the drone strikes. Given the sentiments of Pakistan's Parliament and public, UN special rapporteur Ben Emmerson concluded that, regardless of the continued cooperation of Pakistan's military and intelligence institutions, as a matter of law, "The continued use of remotely piloted aircraft in the Federally Administered Tribal Areas amounts to a violation of Pakistani sovereignty, unless justified under the international law principle of self-defence."[42]

And what about the principle of self-defense? International law allows nations to use force to defend themselves against imminent threat, but in the words of UN rapporteur Christof Heyns, this "may not be done pre-emptively to prevent a threat from arising in the future." Instead, in a much quoted phrase that has survived from a legal case in 1842, the threat of impending attack must be "instant, overwhelming, and leaving no choice of means, and no moment of deliberation."[43] A leaked confidential memo from the Justice Department, however, offers an opinion that seems to be at odds with settled legal precedent—that imminence "does not require the United States to have clear evidence that a specific attack on U.S. persons and interests will take place in the immediate future." The memo says that a "decision-maker determining whether an al-Qaeda operational leader presents an imminent threat of violent attack against the United States must take into account that certain members of al-Qaeda … are continually plotting attacks against the United States; that al-Qaeda would engage in such attacks regularly to the extent it were able to do so; that the U.S. government may not be aware of all al-Qaeda plots as they are developing and thus cannot be confident that none is about to occur."[44] Although the sleight of hand here has been condemned by a number of legal commentators, the pithiest critique comes from the television comedian John Oliver, who drew a vivid analogy in an extended critical commentary on drones: "When someone says, 'I'm going to have a baby imminently,' it doesn't mean, 'I may or may not have a baby at some point in the future.' It means, 'Get your fucking car keys; my water just broke.'"[45]

For the use of military force to be defensible under international law, particularly if it takes place on foreign ground, an attack must be in progress or imminent, and it must be undertaken by an organization recognizable as a military force. (A U.S. embassy might be attacked by a drug cartel, an angry mob, or a motorcycle gang, but no one would say that this constitutes a declaration of or case for war because the attackers would not be seen as a military force.) All three UN special rapporteurs emphasize that, as Ben Emmerson puts it, "individuals can be regarded as members of an armed group, such that they may be targeted for lethal operations at any time, only if they have assumed a continuous combat function within the group. Continuous combat function implies lasting integration into an armed group." He defines an organized armed group as one that has "at least a common command structure, adequate communications, joint mission planning and execution, and cooperation in the acquisition and distribution of weaponry." Emmerson excludes from this definition those who enact "a spontaneous or sporadic or temporary role for the duration of a particular operation."[46] This interpretation of international law is grounded in what we might call the "classic" vision of war as a contest between regular armies whose members are full-time soldiers. This definition of legitimate targets is harder to square with the practice of guerilla war, where fighters do not wear uniforms and may drift in and out of an insurgency, taking up fighting as an occasional or part-time vocation. Guerilla war also is often practiced by decentralized networks of loosely affiliated groups rather than an organization coordinated by a military bureaucracy.

All three UN special rapporteurs conclude that the disparate targets attacked by U.S. drones in various countries scattered around the Middle East fail to rise to the standard of a single armed group with stable personnel with which, under international law, the United States could be said to be at war. Emmerson states that "the United States uses the term 'associated force' as applying to an organized armed group that has entered the fight alongside Al-Qaida and is a co-belligerent with Al-Qaida in the sense that it engages in hostilities against the United States or its coalition partners. There is, however, considerable doubt as to whether the various armed groups operating under the name of Al-Qaida in various parts of the world, or claiming or alleged to be affiliated with Al-Qaida, share an integrated command structure or mount joint military operations."[47] As Philip Alston says of those targeted by U.S. drones, "Sometimes they appear to be not even groups, but a few individuals who take 'inspiration' from al Qaeda. The idea that, instead, they are part of continuing hostilities that spread to new territories as new alliances form or are claimed may be superficially appealing but such 'associates' cannot form a 'party' as required by IHL [international humanitarian law]—although they can be criminals. ... To ignore these minimum requirements," concludes Alston, "would be to undermine IHL safeguards against the use of violence against groups that are not the equivalent of an organized armed group capable of being a party to a conflict."[48]

Some might accuse the UN rapporteurs of engaging in a kind of scholastic sophistry that insists on the immoveable primacy of categories that are in danger of becoming obsolete in the face of the actual organizational practices of Islamist insurgents and their fellow travelers who are attacking Americans. They might say that the United States has the right to defend itself against people with weapons who intend to kill Americans if they can find a way to do so, even if traditional legal categories do not quite fit emerging threats: the categories must bend in light of reality. As General David Deptula put it in an interview with journalist Chris Woods, "There are people out there saying, 'Hold on, this isn't right to use an application of force outside the defined battle area!' Well that's an anachronistic construct that harkens back to the Clausewitzian days where people defined conflict as a result of lines on a map. ... Our adversary isn't limited to lines on a map? Then what is the most effective way to fight that adversary?"[49] This is not the argument that the United States has made, however. Instead, it claims that its drone strikes are clearly justified by the international laws of war because those it targets in Pakistan, Somalia, Syria, and Yemen are part of a single continuous armed threat and because there is in a readily understandable sense a war between the United States and forces in these different countries. The UN rapporteurs—who are tasked with defending the Geneva Conventions, which delimit war as clearly as possible and safeguard the boundaries between combatants and noncombatants—can only be skeptical of such claims. Pointing out that the United States has deliberately killed leaders of drug cartels in Afghanistan with drones,[50] they wonder if the United States does not misclassify as acts of war what should more properly be seen as crimes, thus killing people who should be captured and given a trial rather than killed. Charged with protecting civilians from war and ensuring that, when there is doubt, "the person must be considered a civilian," they believe that the United States has the reverse default, especially when it speaks as if all "military-age males" are legitimate targets. They suspect that, in Christof Heyns's words, "terms such as 'terrorist' or 'militant' are sometimes used to describe people who are in truth protected civilians."[51] (Heyns is particularly concerned that U.S. "double-tap" strikes, which target with a follow-up missile those who rescue the victims of a drone strike, constitute "a war crime in armed conflict and a violation of the right to life.")[52]

Above all, they suspect that the United States is in the position of a person with a hammer who treats everything as a nail: "the greater concern with drones," says Alston, "is that because they make it easier to kill without risk to a State's forces, policy makers and commanders will be tempted to interpret the legal limitations on who can be killed, and under what circumstances, too expansively. States must ensure that the criteria they apply to determine who can be targeted and killed—i.e. who is a lawful combatant and what constitutes 'direct participation in hostilities' that would subject civilians to direct attack—do not differ based on the choice of weapon."[53] Heyns worries that not just the selection of targets but the very bounding of war itself may change as drones become preferred instruments of force: "The ready availability of drones may lead to states … increasingly engaging in low-intensity but drawn-out applications of force that know few geographical or temporal boundaries. This would run counter to the notion that war—and the transnational use of force in general—must be of limited duration and scope, and there should be a time of healing and recovery following conflict."[54] The concern is that the very alterity between war and peace will dissolve as we move to a world of permanent war without demarcated battle zones. This fear that drones will lead to unbounded war also lies at the heart of the final concern expressed by critics that I discuss here—the concern that drones constitute a moral hazard.

Moral Hazard   "Moral hazard" refers to a situation where a person may be willing to take risks because they know someone else will bear the consequences. John Kaag and Sarah Kreps tell us that moral hazard "began as an insurance industry term referring to people's tendency to take greater risks when they do not face the associated costs. Once a car was insured, it was more likely to be driven recklessly."[55]

At the end of the eighteenth century, the philosopher Immanuel Kant applied the logic of moral hazard to the issue of war and peace in his extended essay "Towards Perpetual Peace." He saw that modern technology was making war more expensive and more deadly, and he surmised that democratic states would be less likely to go to war because their governments would be accountable to those who would bear the costs in terms of blood and treasure: "In a constitution which is not republican, and under which the subjects are not citizens, a declaration of war is the easiest thing in the world to decide upon, because war does not require of the ruler, who is the proprietor and not a member of the state, the least sacrifice of the pleasures of his table." On the other hand, "if the consent of the citizens is required in order to decide that war should be declared ... nothing is more natural than that they would be very cautious in commencing such a poor game, decreeing for themselves all the calamities of war."[56]

Kant's claim should be treated with some caution. There are moments when democratic publics hunger for war and a government that stands in their way is a government in peril. An example would be the British public in 1982, when Argentina invaded the Falkland/Malvinas Islands. Still, democratic constitutions do impose consultative procedures on leaders who want to go to war, and a democratically elected government has good reason to fear a war that is going badly and producing high casualties (as Lyndon Johnson's and Richard Nixon's administrations learned during the Vietnam war and the George W. Bush administration learned when wars in Iraq and Afghanistan began to bog down).

The advent of drone warfare changes the calculus here. Drones make it possible to launch attacks in faraway countries without having to mobilize and deploy troops and with no fear of American bodybags. A democratic public can now be as indifferent to the risks of war as an unaccountable sovereign. In the language of economics, drones make it possible to externalize the costs of war. And, as we have seen, because no troops are sent abroad to risk death, the president can insist that he is not engaged in war at all and that the use of drones falls within the sphere of executive privilege. The U.S. experience suggests that, in the face of mass public indifference to the use of drones abroad, few members of Congress will challenge such a claim. In this way, governments may find that drones become preferable to other instruments of military action, and the threshold for military action can be lowered. As Stephen Wrage puts it, "If there are unusually useable weapons in the arsenal, there will be unusual pressures to use them."[57] Thus, Gregoire Chamayou, discussing Kant, says that "as soon as the costs of war become an external matter, the very theoretical model that proclaimed the arrival of a democratic pacifism begins to predict the opposite: a *democratic militarism*."[58]

John Kaag and Sarah Kreps summarize the resulting situation as follows:

> when wars can be fought without young men and women going into battle to kill and be killed, governments do not have to offer an explanation for what they are doing. This will undermine peace and liberal democracies. Ironically, the pressure from a democratic electorate to protect itself from the harms of warfare will not encourage policy makers to adopt peaceful or democratic methods ... but rather methods of warfare that leverage technology to insulate citizen-soldiers from harm. The irony is this insulation creates the possibility that leaders will no longer, in a prudential sense, have to obtain popular permission to go to war.[59]

As Stephen Holmes put it, writing in the *London Review of Books*, "The instrument that has allowed [Obama] to narrow the fight guarantees that the fight will go on."[60]

Drones, Democracy, and War

This chapter has examined what might be described as a debate within liberal theory about war. We have looked at the U.S. government's arguments that drones protect democracy by enabling a discriminate use of force, in tune with liberal and humane values, against the Islamist enemies of democracy and human rights. We have also looked at arguments made by critics of drone warfare that drones enable the circumvention of constitutional checks on a president's ability to take the state to war, and that drones are being used in a way that undermines fundamental precepts in international law that separate combatants from noncombatants while restricting the circumstances in which states can attack the territory of other states. Surveying this debate, one is struck by the legalistic opportunism and disingenuousness with which the Obama administration has made its case, picking and choosing tidbits from the law as if it were a buffet dinner while bulking up its executive power. It is not difficult to see the plausibility of concerns that drones are helping to tilt the system of checks and balances in American government in the president's favor, that they are dissolving the categories that undergird the Geneva Conventions, and that they are making military intervention dangerously cheap and easy.

And yet there is a way in which the debate recounted in this chapter, the debate within liberal theory about war, is somehow askew from the picture it frames. The debate implicitly assumes that there is a stable phenomenon of war that is regulated by a settled apparatus of domestic and international law, and that the U.S. government is now using drones to try and get away with breaking some of the rules about who can be targeted in which circumstances. But it may be fairer to say that the very phenomenon of war, as conceived in liberal international theory, was already in crisis when drones came on the scene. The introduction of drones into the mix has merely intensified a process of categorical collapse that was already well underway, thanks to the emergence of nuclear weapons, strategic bombing, landmines, cluster bombs, death squads, and guerilla movements, none of which aligned well with the niceties of the Geneva Conventions. With the exception of drones that are used alongside ground troops and manned airpower in Iraq and Afghanistan as part of what I earlier called mixed drone warfare, it is not even clear that war is the right framing category for the enterprise in which drones are engaged. This has been obscured by the legions of lawyers who drive the debate on both sides and who argue about drones through the category of war: But remember: "the drone upsets the available categories, to the point of rendering them inapplicable."[61] With the help of the drone, war is fragmenting into something else for which we do not yet have a good name.

In her book *New and Old Wars*, Mary Kaldor notes that the ratio of military to civilian casualties in war in the early twentieth century was eight to two and that by the end of the century it was two to eight.[62] She argues that as the cold war ended and globalization bit deeper, there was increasing fusion between armies, militias, criminal networks, and drug cartels and that war was increasingly defined by the clash of sectarian communities—Catholic versus Protestant, Sunni versus Shi'a, Serb versus Croat, Hutu versus Tutsi—rather than the grand clash of armed states at the heart of classic war. Amid genocides in Rwanda, Yugoslavia, Sri Lanka, and Sudan and following protracted guerilla warfare in Vietnam, Angola, Afghanistan, and Colombia, the boundaries demarcating battlefields and the distinctions between combatants and civilians were in profound crisis before the first Predator drone ever launched a Hellfire missile. The drone's introduction to the mix, if anything, obscured these shifts because it was accompanied by a lawyerly rhetoric that was drawn from the language of the UN charter about the need to discriminate between legal and illegal targets and between combatants and civilians. Even as the drone undermined classic war in practice by playing havoc with the boundaries of the traditional battlefield and creating a more deeply asymmetrical form of warfare, its operations were legitimated by a state rhetoric of discriminate force that seemed to reinstantiate the very legal framework that it was in other ways dissolving.

In asking whether drones can be used as instruments of democratic war, we must question not only whether their use has been compatible with democratic norms but also whether the enterprise in which they have been used can even be called war. Most definitions of war assume that it is a contest in which the act of killing, no matter how asymmetrical, is reciprocal. In chapter 2, I quote the anthropologist Talal Asad, who remarked on "the conventional understanding of war as an activity in which human dying and killing are exchanged."[63] Carl von Clausewitz, seen by many as the greatest theorist of war, begins his work *On War* with this definition: "We shall not enter into any of the abstruse definitions of war used by publicists. We shall keep to the element of the thing itself, to a duel. War is nothing but a duel on an extensive scale."[64] Likewise, in her book *The Body in Pain*, which compares the grammatical structures of war and torture, Elaine Scarry titles one section "War Is a Contest." She says that in this contest, "the participants must work to out-injure each other. Although both sides inflict injuries, the side that inflicts the greater injury faster will be the winner; or, to phrase it the other way round, the side that is more massively injured or believes itself to be so will be the loser."[65] She argues that it is not war when one side inflicts injury on the other without the possibility of reciprocal injury; it is then torture. By its nature, war is structured around the reciprocal infliction of pain and death. It is a contest, maybe an uneven one, but a contest nonetheless, and either side can "score" in principle. In torture, which is about domination rather than contestation, pain is inflicted unilaterally. There is no mystery as to who will be hurt; the only question is whether the victims will betray their cause. Scarry was writing at the height of the nuclear arms race and argued that nuclear weapons were too destructive to be contained within an injuring contest and that "nuclear war" was therefore a grammatical impossibility. Her theoretical frame offers a suggestive way of looking at drone operations, which, like torture, involve a unilateral infliction of pain. One implication of Scarry's work is that the term "drone warfare" is, in the absence of any opportunity for reciprocal injuring, a contradiction in terms.

But if drones are not engaged in warfare, what should we call the enterprise in which they are involved? In *A Theory of the Drone*, Gregoire Chamayou gives various answers to this question. Commenting that "warfare, from being possibly asymmetrical, becomes absolutely unilateral," he says that "what could still claim to be combat is converted into a campaign of what is, quite simply, slaughter."[66] Later, he says that with drones "war degenerates into a putting-to-death" and that the new U.S. military ethic is "an ethic for butchers or executioners, but not for combatants."[67] Elsewhere in the book, ceding a little more agency to the victims, he prefers the metaphor of hunting: "contrary to Carl von Clausewitz's classical definition, the fundamental structure of this type of warfare is no longer that of a duel, of two fighters facing each other. The paradigm is quite different: a hunter advancing on a prey that flees or hides from him."[68] Still elsewhere, Chamayou says that "a single decade has seen the establishment of an unconventional form of state violence that combines the disparate characteristics of warfare and policing without really corresponding to either."[69] The Obama administration has sometimes preferred the vocabulary of policing to that of warfare when defending drone operations; but it is an odd kind of police work that can kill and never capture the criminals it pursues and that administers a premeditated death sentence without any recognizable judicial procedure. Finally, at one point Chamayou quotes Major-General Charles Dunlap of the U.S Air Force as making the remarkable statement that "American precision airpower is analogous (on a much larger and more effective scale) to the effect that insurgents try to impose ... through their use of improvised explosive devices" and that the purpose of "precision airpower" is to inculcate a "hopelessness that arises from the inevitability of death from a source they cannot fight" so as to "extinguish the will to fight." This leads Chamayou to suggest that "drone strikes are equivalent to bomb attacks. They constitute weapons of state terrorism."[70] Although this might seem like strong language, one can readily imagine that it would be called terrorism if a foreign government killed one of its enemies with a car bomb on U.S. soil, even if it ensured that no civilians were killed or hurt.

Whatever we call what drones do—slaughter, hunting, aerial police work, targeted killing, state terrorism, warfare—these machines and their operators are remaking the world in significant ways. They are enabling a kind of permanent, low-level military action that threatens to erase the boundary between war and peace and, in its departure from classic war, is not easily contained or regulated by either the War Powers Act or the UN Charter. Unlike earlier technologies (such as nuclear weapons and landmines) that undermined the Geneva Conventions, drones do not kill indiscriminately. They can be highly discriminate in their use, and thus one can see their appeal to human rights hawks. But, as is clear from the testimony of Pakistanis who have the misfortune to live under the drones, they are also terror weapons. Above all, they are reconfiguring the relationship between military power, national sovereignty, and territorial boundaries in ways that are in profound tension with existing international law. In Chamayou's words, "What is emerging is the idea of an invasive power based not so much on the rights of conquest as on the rights of pursuit: a right of universal intrusion or encroachment that would authorize charging after the prey wherever it found refuge, thereby trampling underfoot the principal [sic] of territorial integrity classically attached to state sovereignty. According to such a concept, the sovereignty of other states becomes a contingent matter."[71]

We must correct Chamayou in one regard, however. The emerging right of pursuit is in no way universal. The United States would never dream of using a drone to kill someone in France, Germany, or Russia. Julian Assange and Edward Snowden are safe in the United Kingdom and Russia, respectively, in a way that Anwar al-Awlaki never could be in Yemen. Drones are an imperial border-control technology for the age of late capitalism. They can be used only against countries that lack the technological sophistication to shoot down the slow-moving planes and whose internal affairs, conforming to Western stereotypes of "failed states," provide a pretext for incursion that is as persuasive to liberal interventionists today as the white man's burden was to their Victorian ancestors. The drone strike is represented as a means of upholding a universal international law but is, in practice, a means of stratifying global society. Enacted on the basis that countries such as Yemen and Somalia are in some sense lawless, the drone strike further establishes through its performance the abject status of countries that do not have the legal right to territorial integrity enjoyed by more mature nations. As the drone undermines the territorial boundaries of some countries, demonstrating that they can be subject to repeated microinvasions from the air, it simultaneously bifurcates the world, establishing a moveable master seam between one side where drones can be used and another side where they cannot. Reliant on the communications infrastructure that only the most advanced industrial societies have developed, but too vulnerable to use outside the least militarily developed countries on earth, the drone is an inherently colonialist technology that makes it easier for the United States to engage in casualty-free, and therefore debate-free, intervention while further militarizing the relationship between the U.S. and the Muslim world.

Conclusion: Peering over the Horizon

I think of where the airplane was at the start of World War I: at first it was unarmed and limited to a handful of countries. ... Then it was armed and everywhere. That is the path we're on.

—Peter Singer[1]

Peering over time's horizon, one can easily imagine two different futures. In one scenario, a dystopian future, drones are used to kill anyone from insurgents to criminals and dissidents. In this future, U.S. drones kill insurgents, drug cartel leaders, and people who hack into corporate bank accounts in Mali, the Central African Republic, Venezuela, Colombia, the Philippines, Yemen, and Somalia. The United States also uses a fleet of drones to patrol the U.S.-Mexico border. Congress, worried that drones might kill people within the United States, initially requires drones to be used only for surveillance. But after undocumented Mexican men connected to a drug cartel murder a married couple and their two young children in Arizona, the White House wins the right to fit drones with "nonlethal weapons," such as airborne tasers, tear gas, and guns with rubber bullets. The rules of engagement for deploying these weapons loosen over time, and adults and children crossing the border illegally are killed when tasers cause heart attacks and rubber bullets pierce eye sockets.

In this dystopian future, other countries also have deployed their own fleets of drones.[2] Russia, following less stringent rules of engagement, uses drones to hunt and kill Chechens and Ukrainian nationalists. Israel continues to use drones to kill Hamas leaders in Gaza. China uses drones against the Uighur rebellion that has been simmering for over a decade and starts to use drones to kill Tibetan leaders whom it accuses of terrorist plots. Even middle-tier countries rent satellite capability from China and from private corporations in the United States so that they can deploy their own drones. Nigeria has two drones that it uses against Muslim separatists in its north, Turkey uses drones against a revived Kurdish rebellion, Pakistan uses its own drones against separatists in Waziristan,[3] and experts in international security fear that India's drones patrolling the border with Pakistan could become a flashpoint for a new war between those two countries.

In another facet of this dystopian future, the United States has developed a new generation of kamikaze assassination micro (KAM) drones. These small devices are easily confused with large insects, and they use facial recognition software to identify individual targets and kill them. Some KAM drones do this by exploding in the victim's face, and others, equipped with sharp blades, plunge themselves at high speed into the neck of the victim, severing an artery as they strike.[4] The earliest KAM drones relay photographs back to controllers in Northern Virginia, who confirm that the victim matches a facial database and then approve the strike. After a few months, however, the CIA director persuades the president that the automated facial recognition software has a very low rate of false positives (an investigative article in the *New York Times* several years later casts doubt on this claim). The KAM drones are made completely autonomous after a man who bombs an American consulate escapes when a contractor in Northern Virginia, distracted by an online video game he is playing, takes too long to approve the match. Because President Gerald Ford's prohibition on assassination had been abandoned years earlier, the CIA also persuades the president that although assassination is held in low regard, it is better to kill individual leaders in foreign countries than to engage in wars in which many more people will die. The CIA is surprised by the speed with which Mexican drug cartels procure their own assassination micro drones because the agency thought that the technology was too sophisticated for them to master.

In this scenario, drones also become an important technology for police departments. Unlike earlier times, when police officers used to walk, bicycle, or drive through neighborhoods, getting to know the people they policed, they increasingly sit in windowless trailers, scanning for criminals from above. Police drones are not supposed to be weaponized, but exceptions have been made in the case of street riots in Baltimore and Cleveland. Although the American Civil Liberties Union and other rights groups have complained that drones violate the Fifth Amendment, particularly when they use infrared cameras to look inside houses and apartments, the courts have rejected their arguments because of the public's concerns about terrorism and gang crimes.

Nor is the fear of terrorism ill-founded: an array of groups from right-wing anti-government militia to Islamist cells have acquired their own crude drone capabilities. An Islamist group buys a $500 drone online, flies the drone into a 737, and brings down the aircraft on takeoff.[5] The group's statement says, "The infidels who have killed our women and children from the air have now died in the air at the hands of their own cowardly technology. More attacks will follow." At the same time a group calling itself the McVeigh Brigade fits several drones with crude explosives and flies them into a shopping mall, where they kill forty-three people, including twelve teenagers. And a number of individuals create crude weaponized drones by attaching hand-guns to recreational drones bought off the shelf and use them to kill spouses, neighbors, and coworkers.[6] Where mass shootings used to be done face-to-face, they are now increasingly done by drone.

In a second, alternative, scenario for the future, drone use is more regulated and benign. Drone technology has many benefi-cial uses (of which the possibility of speedier delivery by Amazon is among the most publicized and the least important). At the time of writing, drones are already being used to hunt for poach-ers in Africa, map environmental damage, guide firefighters in national forests, monitor the site of the Fukushima accident without exposing humans to radiation, search for lost hikers in the wilderness, replace humans in the dangerous work of inspect-ing power lines, and so on. So the alternative scenario for the future is one in which drones exist, they are well regulated by both international and domestic laws, and their civilian uses outweigh military applications. In this other version of the future, drones are part of the arsenal of war alongside attack heli-copters and manned planes, in formally recognized warzones, but their use for targeted killing outside formal warzones is strictly banned—just as it would be banned for, say, Venezuela to use a car bomb to kill an opposition leader in Washington DC, even if the car bomb killed no-one but the intended target, and even if there was reason to believe that the target had been responsible for the deaths of Venezuelan civilians in the past. In this version of the future, an international treaty requires that a human be "in the loop" when a decision is made to kill someone so that machines do not make automated decisions to kill.[7] The same treaty requires that all drone video footage and the discussions leading to drone strikes be archived so that they can be reviewed by an international court for evidence of criminal negligence or culpability in the event that a drone strike kills civilians.

In this alternative future, Congress has responded to concerns that drone surveillance undermines a fundamental right to privacy and has limited the use of drones within the United States by domestic law enforcement agencies. Drones flying over the United States cannot be weaponized, their surveillance is limited to public spaces, and the video footage that they collect must be deleted within twenty-four hours unless a judge grants an exemption. Congress has passed a law against "drone stalking" and "peeping drones," making it illegal for private citizens (including mistrustful spouses) to track or spy on people with drones.[8]

In this second scenario, commercial drones are regulated so that they do not interfere with the operations of commercial airlines. Commercial and amateur drone operators have to pass driving tests, there are strict penalties for flying a drone while drunk, and all drones are chipped and registered in a national database to ensure liability in case of accidents. Hobbyist drones are fitted with chips that prevent them from flying above a certain altitude, and geofencing protects sensitive spaces (sports stadiums and the White House grounds, for example) from intrusions by drones.

Most people will find the second scenario for the future more attractive than the first. It is also the less likely of the two. The first, dystopian scenario assumes that drone technology (like other military technologies) will be adopted by other countries and even nonstate actors such as terrorists and drug cartels and that scientists will produce new iterations of the technology that will make drones smaller, deadlier, and cheaper. It also assumes that the rules of engagement will loosen (through a process of slippage analogous to that documented in chapter 4 of this book) so that drones will be used in increasingly permissive ways. (However, chapter 4 also shows that the process of slippage in the case of U.S. drone strikes against Pakistan was eventually reversed, so we have empirical evidence that it is possible to tighten the chain on drone use.)

So the question is: how do we control this new technology? How do we go about drone arms control?

Some activists have suggested that military drones might be banned in the way landmines and cluster bombs were.[9] Such a suggestion strikes this author as impractical and even undesirable. The same drone can be used to perform a targeted killing or to verify compliance with an arms control treaty. Surely we do not want to lose the second function. UN Secretary-General Ban Ki-moon has suggested a treaty that allows drones for surveillance but bans weaponized drones,[10] but this is problematic for three reasons. First, a drone that is "painting" a target with a laser so that a neighboring manned plane can destroy the target can be characterized as both a weaponized drone and a surveillance drone. Second, a plane should not be banned simply because the pilot flies it remotely rather than from a cockpit inside the plane. What matters is the effects of the weapon, not the location of its operator. Third, in principle drones can be used to strike targets on the ground more discriminiately than manned planes, because they can gather extensive intelligence before an attack and stream live video footage to multiple control centers when an attack is being considered. In chapter 1, I distinguished between what I called "mixed drone warfare" (when drones are used in concert with ground forces and manned aircraft as part of a larger war effort) and "pure drone warfare" (when drones are used as standalone weapons to kill people without warning in contexts where the United States is not publicly at war). If weaponized drones were banned, we would be saying that people in Afghanistan could be attacked with F-16s, Apache helicopters, and artillery but not with drones, which are more likely to spare innocent civilians. That would surely be perverse. The problem is not the weaponized drone in all contexts but the weaponized drone used as a tool of assassination outside the permitted context of war as well as relaxed targeting protocols that do not hold drone operators to sufficiently high standards in discriminating combatants from civilians. (This is, until September 2015 at least, why the United Kingdom used its drones to attack targets in Afghanistan but refused to deploy them alongside U.S. drones against targets in other countries.) In chapter 5, I quote Harold Koh, a senior legal adviser in the Obama administration, who critiques what I call "drone essentialism" by saying that "it makes as little sense to attack drone technology as it does to attack the technology of such new weapons as spears, catapults, or guided missiles in their time. Cutting-edge technologies are often deployed for military purposes; whether or not that is lawful depends on whether they are deployed consistently with the laws of war, *jus ad bellum* and *jus in bello*."[11] Following Koh's logic, we can say that it makes more sense to police the uses to which drones are put than to attempt to ban the technology outright. That would mean undoing the damage that Koh himself did to international law and establishing that drones (like any other form of airpower) can only be used against targets in legally recognized war zones and that it is as illegal to kill someone outside a formal warzone with a Hellfire missile launched from a drone as it would be to walk up to them with a gun and shoot them in the head in another sovereign nation with which one is not at war.

U.S. leaders like drones, however, precisely because they can be used outside formal war zones to kill people in places like Yemen and Somalia or the so-called tribal areas of Pakistan without putting U.S. troops in danger. If we want to restrict the use of weaponized drones to internationally recognized war zones and ban their use as weapons of assassination and terror on the frontiers of empire we can expect little help from the U.S. government. If targeted killing outside the law has been so attractive to a president who was a constitutional law professor, who opposed the war in Iraq from the very beginning, who ended the Central Intelligence Agency's torture program, and who announced his intention to close the Guantanamo Bay detention camp on assuming office, it is unlikely that any successor to his office will easily renounce the seductions of the drone. At the same time, the machinery of checks and balances installed by the founding fathers has seized up in relation to drones and has effectively handed the president a blank check to do as he pleases.

If targeted killing on the frontier is to be banned, the impetus will likely come from outside the United States—just as it did with the treaties banning landmines and cluster bombs (neither of which has been signed by the United States). This could happen through a transnational process, led by nongovernmental organizations and sympathetic states, of the kind that produced the bans on landmines and cluster bombs. It might conceivably be imposed by a World Court decision, although the United States has a history of ignoring World Court decisions that it dislikes. Or it could happen in response to a court decision in another country to arrest a U.S. official who is responsible for ordering drone strikes on war crimes charges, or to levy fines in compensation for depriving people of their right to life.

There is one more possibility. In recent years, the United States has stopped sending people that it suspects of being insurgent leaders to Guantanamo Bay and "black sites" where they were tortured. This shift took place because those responsible for the program feared possible criminal charges in the future, and because they came to realize that this approach to counterinsurgency was not producing the intelligence information that had been foreseen but was blackening the international reputation of the United States. (The approach that was substituted was to rely on drones to kill those who formerly would have been captured and interrogated.)[12] A similar development can be imagined with regard to drones, which have hurt America's reputation all over the world, but especially in the Middle East. The government's claims about the ability of drones to discriminate civilians from militants has only increased public anger when civilians are killed. In the words of Stephen Holmes,

The rage such strikes incite will be all the greater if onlookers believe, as seems likely, that the killing they observe makes relatively little contribution to the safety of Americans. Indeed, this is already happening, which is the reason that the drone, whatever its moral superiority to land armies and heavy weaponry, has replaced Guantánamo as the incendiary symbol of America's indecent callousness towards the world's Muslims. As Bush was the Guantánamo president, so Obama is the drone president. This switch, whatever Obama hoped, represents a worsening not an improvement of America's image in the world.[13]

Although it is becoming clearer that drones impose a significant public relations cost on the United States, it is not clear that there is a balancing strategic gain. Liquidating people on a kill list is not the same as defeating an insurgency, especially if the kill list seems to grow faster than the people on it can be killed—or indeed to grow in angry response to the impersonal killing from the air of those already on it. Nor do drones, hovering 25,000 feet above the insurgency, enable the painstaking accumulation of human intelligence and cultural understanding, or the patient building of relationships between imperial administrators and native collaborators, that the savviest apostles of counterinsurgency identify as the key to success.[14] Maybe the United States will rein in drone warfare not because it is at cross-purposes with international law, or because it moves us closer to a state of permanent war outside the regulatory reach of democratic institutions, but because its military and intelligence leaders concluded that it does not work.

Notes

Chapter 1: Drones 101

1. "Generation Kill: A Conversation with Stanley McChrystal," *Foreign Affairs*, March/April 2013.

2. Greg Miller, "Brennan Speech Is First Obama Acknowledgement of U.S. Use of Armed Drones," *Washington Post*, April 30, 2012.

3. Greg Miller, "Panetta Clashed with Agency over Memoir, Tested Agency Review Process," *Washington Post*, October 21, 2014.

4. Vali Nasr, "The Inside Story of How the White House Let Diplomacy Fail in Afghanistan," *Foreign Policy*, March 4, 2013.

5. Steven Coll, "The Unblinking Stare," *New Yorker*, November 24, 2014.

6. Matt J. Martin and Charles W. Sasser, *Predator: The Remote-Control Air War over Iraq and Afghanistan* (Minneapolis, MN: Zenith Press, 2010), 18.

7. Jane Mayer, "The Predator War," *New Yorker*, October 26, 2009.

8. Karen McVeigh, "Drone Strikes; Tears in Congress as Pakistani Family Tells of Mother's Death," *The Guardian*, October 29, 2013. The Democratic Congressman from Florida who invited the testimony was Alan Grayson.

9. Emphasizing the importance of other sensory modes than the visual for the victims of drone strikes, the Palestinian artist Rehab Nezzal has

created an exhibit consisting only of the sounds made by drones. See https://www.imagesfestival.com/calendar.php?exhibition_id=297.

10. Gregoire Chamayou, *A Theory of the Drone* (New York: New Press, 2013), 32.

11. http://www.oed.com/search?searchType=dictionary&q=drone &_searchBtn=Search.

12. See "Radio Controls Robot Plane on Pilotless Flight," *Popular Mechanics*, October 1935, p.551. See also http://www.captainnevillesflyingcircus .org.uk/page16.htm.

13. Adam Rothstein, *Drone* (London: Bloomsbury Books, 2015), 27; Medea Benjamin, *Drone Warfare: Killing by Remote Control* (New York: OR Books, 2012), 13; Chamayou, *A Theory of the Drone*, 26.; John Sifton, "A Brief History of Drones," *The Nation*, February 27, 2012; Katherine Chandler, "Drone Flights and Failures," paper presented at Society for Social Studies of Science, San Diego, October 9–12, 2013. See also Katherine Chandler, "Drone Flight and Failure: the United States' Secret Trials, Experiments, and Operations in Unmanning, 1936–1973," PhD diss., University of California, Berkeley, 2014.

14. Lone Wolf Media, *The Bomb*, PBS 2015, http://www.pbs.org/program/ bomb.

15. Rothstein, *Drone*, 28–30.

16. Chamayou, *A Theory of the Drone*, 27; Rothstein, *Drone*, 31.

17. Woods, *Sudden Justice*, 30.

18. Loren Thompson, quoted in Tom Vanden Brook, "Drone Attacks Hit High in Iraq," *USA Today*, April 29, 2008. See also Richard Whittle, "The Man Who Invented the Predator," *Air and Space Magazine*, April 2013; Richard Whittle, *Predator: The Secret Origins of the Drone Revolution* (New York: Henry Holt, 2014); Williams, *Predators*, 20; Chamayou, *A Theory of the Drone*, 29; Benjamin, *Drone Warfare*, 14–16; Andrew Cockburn, *Kill Chain: The Rise of the High-Tech Assassins* (New York: Henry Holt 2015), chap. 4.

19. Fariborz Haghshenass, *Iran's Asymmetric Naval Warfare*, Washington Institute for Near East Policy, Policy Focus 87, September 2008, 17.

20. Woods, *Sudden Justice*, 40.

21. Williams, *Predators*, 24.

22. Woods, *Sudden Justice*, 24–25. And see National Commission on Terrorist Attacks, *9/11 Commission Report* (Washington, DC: Government Publishing Office, 2004), 210–214.

23. Joel Greenberg, "Israel Affirms Policy of Assassinating Militants," *New York Times*, July 5, 2001.

24. Cofer Black, testimony before the U.S. Senate, September 26, 2002, archived at http://fas.org/irp/congress/2002_hr/092602black.html. Woods, *Sudden Justice*, xv.

25. Coll, "The Unblinking Stare."

26. Woods, *Sudden Justice*, 23–26.

27. Quoted in Peter Singer, *Wired for War: The Robotics Revolution and Conflict in the Twenty-first Century* (New York: Penguin, 2009), 397; Doug Struck, "Casualties of U.S Miscalculations: Afghan Victims of C.I.A. Missile Strike Decried as Peasants, Not Al Qaeda," *Washington Post*, February 11, 2002.

28. Alice Ross, "Who Is Dying in Afghanistan's 1,000-Plus Drone Strikes?," July 24, 2014, https://www.thebureauinvestigates.com/2014/07/24/who-is-dying-in-afghanistans-1000-plus-drone-strikes; Associated Press, "US Drone Strikes in Afghanistan Rose Sharply Last Year, UN Reports," *The Guardian*, February 19, 2013; Chris Woods and Alice Ross, "Revealed: U.S. and Britain Launched 1,200 Drone Strikes in Recent Wars," December 4, 2012, http://www.thebureauinvestigates.com/2012/12/04/revealed-us-and-britain-launched-1200-drone-strikes-in-recent-wars.

29. Woods and Ross, "Revealed."

30. Greg Miller, "U.S. Launches Secret Drone Campaign to Hunt Islamic State Leaders in Syria," *Washington Post*, September 2, 2015, A1.

31. Ibid.

32. http://costsofwar.org/article/us-killed-0. These numbers do not include private contractors killed.

33. CNN, "U.S. Kills Cole Suspect," *CNN.com*, November 5, 2002; Greg Miller and Josh Meyer, "CIA Strike May Link Six in U.S. to Al Qaeda," *Los Angeles Times*, November 9, 2002; Williams, *Predators*, 43. The American killed in the strike is named by some sources as Ahmad al-Hijazi and by others as Kemal Darwish. The CIA was aware that the drone strike would kill an American citizen according to Woods, *Sudden Justice*, 58–59, and Dana Priest, "U.S. Military Teams, Intelligence Deeply Involved in Aiding Yemen on Strikes," *Washington Post* January 26, 2010.

34. Woods, *Sudden Justice*, 57.

35. Quoted in Bootie Cosgrove-Mather, "Remote-Controlled Spy Planes," *CBS News*, November 6, 2002.

36. Woods, 64.

37. Ibid., 60–61.

38. Scott Shane, "Documents on 2012 Drone Strike Detail How Terrorists Are Targeted," *New York Times*, June 24, 2015, A11.

39. Jack Serle, "What Next for Yemen as Death Toll from Confirmed U.S. Drone Strikes Hits 424, Including Eight Children," *Bureau of Investigative Journalism*, January 30, 2015; "Drone Wars Yemen: Analysis," New America Foundation, http://securitydata.newamerica.net/drones/yemen/analysis.html; Jeremy Scahill, "Find, Fix, Finish," *The Intercept*, October 15, 2015; for a CNN chart showing the frequency of drone strikes in Yemen, see http://i2.cdn.turner.com/cnnnext/dam/assets/130107043452-chart-drone-strikes-pakistan-yemen-story-top.jpg.

40. Drones Team, "Somalia: Reported U.S. Covert Actions, 2001–2015," *Bureau of Investigative Journalism*, February 22, 2015; Jack Serle, "Does Latest Drone Strike on Al Shabaab Signal Change in US Tactics in Somalia?," *Bureau of Investigative Journalism*, February 6, 2015.

41. Dylan Matthews, "Everything You Need to Know about the Drone Debate, in One FAQ," *Washington Post*, March 8, 2013; Woods and Ross, "Revealed."

42. Mark Mazzetti, "The Drone Zone," *New York Times Magazine*, July 8, 2012, MM32; Akbar Ahmed and Frankie Martin, "Deadly Drone Strikes on Muslims in the Southern Philippines," Brookings Institution, March 5, 2012.

43. Coll, "The Unblinking Stare."

44. James Cavallaro, Stephan Sonnenberg, and Sarah Knuckey, *Living under Drones: Death, Injury, and Trauma to Civilians from US Drone Practices in Pakistan* (Stanford, CA: International Human Rights and Conflict Resolution Clinic, Stanford Law School, and Global Justice Clinic, New York University School of Law, 2012), 25.

45. https://understandingempire.files.wordpress.com/2013/03/drone -strikes-table1.png.

46. http://apps.washingtonpost.com/foreign/drones/#.

47. Williams, *Predators*, 55–60.

48. Madiha Afzal, "Drone Strikes and Anti-Americanism in Pakistan," Brookings Institution, February 7, 2013; Pew Research Center, "Pakistani Public Opinion Ever More Critical of U.S.," June 27, 2012.

49. Woods, *Sudden Justice*, 2, 42.

50. Colin Freeman, "Armchair Killers: Life as a Drone Pilot," *Daily Telegraph*, April 11, 2015.

51. Singer, *Wired for War*, 32–35; http://fas.org/irp/program/collect/ predator.htm; Martin and Sasser, *Predator*, 18–22; Jeremiah Gertler, *U.S. Unmanned Aerial Systems* (Washington DC: Congressional Research Service Report, 7–5700, 2012), http://fas.org/sgp/crs/natsec/R42136.pdf.

52. Coll, "The Unblinking Stare"; Williams, *Predators*, 70–71; http:// www.globalsecurity.org/military/systems/aircraft/mq-9.htm; Gertler, *U.S. Unmanned Aerial Systems*, 35–36.

53. Winslow Wheeler, "The MQ-9's Cost and Performance," *Time*, February 28, 2012.

54. Mica Zenko, "Ten Things You Didn't Know about Drones," *Foreign Policy*, February 27, 2012.

55. Cockburn, *Kill Chain*, chap. 10.

56. Quoted in Chamayou, *A Theory of the Drone*, 12.

57. Jeremy Scahill, Find, Fix, Finish," *The Intercept*, October 15, 2015.

58. Stephen Holmes, "What's in It for Obama?," *London Review of Books* 35, no. 14 (July 18, 2013): 15–18.

59. Wheeler, "The MQ-9's Costs," makes the case that drone expenses have been underestimated. His argument is dismantled in James Hasik, "Affordably Unmanned," June 20, 2012, http://www.jameshasik.com/weblog/2012/06/affordably-unmanned-a-cost-comparison-of-the-mq-9-to-the-f-16-and-a-10-and-a-response-to-winslow-whe.html. On the F-35, see David Francis, "How the DoD's $1.5 Trillion Jet Broke the Air Force," *CNBC.com*, July 31, 2014.

60. "Flight of the Drones," *The Economist*, October 8, 2011.

61. Craig Whitlock, "When Drones Fall from the Sky," *Washington Post*, June 20, 2014, A1. By 2014, there had been 194 accidents that either destroyed the drone or caused more than $2 million in damage. A quarter of these accidents took place in the United States. Another 224 drones were involved in accidents that cost between $500,000 and $2 million. Some of the crashes have been attributed to a muffler that was installed at the insistence of the CIA and that has caused engine overheating. Cockburn, *Kill Chain*, 55.

62. Scott Shane, "Drone Strikes Reveal Uncomfortable Truth: U.S. Is Often Unsure about Who Will Die," *New York Times*, April 24, 2015, A1.

63. Quoted in Cockburn, *Kill Chain*, 117.

64. Dana Priest and William Arkin, "'Top Secret America': A Look at the Military's Joint Special Operations Command," *Washington Post*, September 2, 2011.

65. Robert H. Scales, "The Only Way to Defeat the Islamic State," *Washington Post*, September 7, 2014, A19.

66. The Associated Press reported that documents captured from insurgents in Mali discussed ways of tricking drones by using mannequins to stage fake gatherings, using camouflage mats, and other measures. See Asawin Suebseng, "Drones: Everything You Ever Wanted to Know But Were Afraid to Ask," *Mother Jones*, March 5, 2013.

67. Cockburn, *Kill Chain*, 256.

68. John Reed, "Predator Drones 'Useless' in Most Wars, Top Air Force General Says," *Foreign Policy*, September 19, 2013.

69. For a partial list of drone crashes, see http://dronewars.net/drone -crash-database.

70. Eyal Weizman, *Hollow Land: Israel's Architecture of Occupation* (New York: Verso, 2013), 239. On the American faith in air power, including a discussion of Douhet's foundational role, see Michael Sherry, *The Rise of American Air Power: The Creation of Armageddon* (New Haven, CT: Yale University Press, 1987).

71. Shane Riza, *Killing without Heart* (Washington, DC: Potomac Books, 2013), 53.

Chapter 2: War Remixed

1. George Brant, *Grounded* (London: Oberon Books, 2013), 38.

2. Matt J. Martin and Charles W. Sasser, *Predator: The Remote-Control Air War over Iraq and Afghanistan: A Pilot's Story* (Minneapolis, MN: Zenith Press, 2010).

3. Micah Zenko, "Ten Things You Didn't Know about Drones," *Foreign Policy*, February 27, 2012. Different sources give slightly different

numbers of people needed to keep a drone aloft. For example, *The Economist* says the number is 180. See "Flight of the Drones," *The Economist*, October 8, 2011.

4. Craig Whitlock, "U.S. Drone Base in Ethiopia Is Operational," *Washington Post*, October 27, 2011; Craig Whitlock, "Remote U.S. Base at Core of Secret Operations," *Washington Post*, October 25, 2012; Craig Whitlock, "Pentagon Set to Open Second Drone Base in Niger as It Expands Operations in Africa," *Washington Post*, September 1, 2014; Greg Jaffe, "Fleet of U.S. Drones Now Based in Turkey," *Washington Post*, November 14, 2011; Greg Miller, "U.S. Launches Secret Drone Campaign to Hunt Islamic State Leaders in Syria," *Washington Post*, September 2, 2015; Nick Turse, "Target Africa," *The Intercept*, October 15, 2015; Brian Glyn Williams, *Predators: The CIA's Drone War on El Qaeda* (Dulles, VA: Potomac Books, 2013); Chris Woods, *Sudden Justice: America's Secret Drone Wars* (New York: Oxford University Press, 2015), 39.

5. Martin and Sasser, *Predator*, 245.

6. Ibid.; see also 23–25.

7. Craig Whitlock, "When Drones Fall from the Sky," *Washington Post*, June 20, 2014.

8. Elijah Solomon Hurwitz, "Drone Pilots: 'Overworked, Underpaid, and Bored,'" *Mother Jones*, June 18, 2013.

9. Robert Kaplan, "Hunting the Taliban in Las Vegas," *Atlantic Monthly*, September 2006, 81–84.

10. Matthew Power, "Confessions of a Drone Warrior," *GQ*, October 23, 2013, 192.

11. Zenko, "Ten Things You Didn't Know"; Ellen Nakashima and Craig Whitlock, "With Air Force's Gorgon Drone, 'We Can See Everything,'" *Washington Post*, January 2, 2011; Woods, *Sudden Justice*, 10–11.

12. Martin and Sasser, *Predator*, 30.

13. Power, "Confessions of a Drone Warrior."

14. David Wood, "Drone Strikes: A Candid, Chilling Conversation with Top U.S. Drone Pilot," *Huffington Post*, May 15, 2013.

15. Peter Singer, *Wired for War: The Robotics Revolution and Conflict in the Twenty-first Century* (New York: Penguin Press, 2009), 366.

16. Andrew Cockburn, *Kill Chain: The Rise of the High-Tech Assassins* (New York: Henry Holt, 2015), 174–175.

17. Singer, *Wired for War*, 351–352.

18. Martin and Sasser, *Predator*, 222. *Stove-piping* (also called *siloing*) refers to the practice, common in large bureaucratic organizations, of creating separate information flows that make it hard for decision makers to see the overall picture.

19. Robert Koehler, "'Bugsplat': The Civilian Toll of War," *Baltimore Sun*, January 1, 2012.

20. Martin and Sasser, *Predator*, 53.

21. Power, "Confessions of a Drone Warrior."

22. Cockburn, *Kill Chain*, 5.

23. Martin and Sasser, *Predator*, 51–54.

24. David Rohde, "The Drone Wars," *Reuters Magazine*, January 26, 2012; Bill Moyers and Michael Winship, "Hubris of the Drones," http://readersupportednews.org/opinion2/277-75/16015-the-hubris-of -the-drones, February 13, 2013.

25. Justin Randle, "Low-Flying Drones," *London Review of Books*, blog, March 20, 2013.

26. Brant. *Grounded*, 42.

27. Williams, *Predators*, 1; Akbar Ahmed, *The Thistle and the Drone: How America's War on Terror Became a Global War on Tribal Islam* (Washington, DC: Brookings Institution Press, 2013), 297.

28. Ahmed, *The Thistle and the Drone*, 298.

29. Usama Khilji, "Living under Drones," *Daily Times* (Pakistan), May 10, 2012, quoted in Ahmed, *The Thistle and the Drone*, 83.

30. Quoted in James Cavallaro, Stephan Sonnenberg, and Sarah Knuckey, *Living under Drones: Death, Injury, and Trauma to Civilians from U.S. Drone Practices in Pakistan* (Stanford, CA: International Human Rights and Conflict Resolution Clinic, Stanford Law School, and Global Justice Clinic, New York University School of Law, 2012), 82.

31. Steven Coll, "The Unblinking Stare," *The New Yorker*, November 24, 2014, 98.

32. Gregoire Chamayou, *A Theory of the Drone* (New York: New Press, 2013), 45.

33. Cavallaro, Sonnenberg, and Knuckey, *Living under Drones*, 80–88. These symptoms resemble those experienced by Guatemalan villagers who live near death squads, as described by the anthropologist Linda Green in her book *Fear as a Way of Life* (New York: Columbia University Press, 1999).

34. Cavallaro, Sonnenberg, and Knuckey, *Living under Drones*, vii.

35. For a drone-themed Pakistani art project, see http://www.motherjones.com/media/2013/06/pakistani-drone-art-mahwish-chishty.

36. Alan Grayson, "Has It Become Too Easy to Kill?," *Reader Supported News*, November 7, 2013.

37. The lyrics of Waters's song "The Bravery of Being out of Range" can be found at http://www.metrolyrics.com/the-bravery-of-being-out-of-range-lyrics-roger-waters.html.

38. Chamayou, *A Theory of the Drone*, 114, 118.

39. To complicate this narrative a little, there is a long history in war of using booby traps (of which the improvised explosive device is the latest example), and booby traps also disarticulate in space the body and weapon of the attacker. However, booby traps and landmines have been largely passive technologies activated fortuitously by the enemy,

whereas drones enable the attacker to act while being remote from the scene of the attack.

40. Quoted in Stephen Holmes, "What's in It for Obama?," *London Review of Books* 35, no. 14, July 8, 2013, 15–18. Richard Clarke is apparently unaware that some drone operators are female.

41. Woods, *Sudden Justice*, 41.

42. For an example of such commentary, see Glenn Greewald, "Bravery and Drone Pilots," *Salon*, July 10, 2012.

43. Chamayou, *A Theory of the Drone*, 111.

44. My thanks to one of the anonymous reviewers for this point.

45. Chamayou, *A Theory of the Drone*, 120.

46. Brant, *Grounded*, 36, 51. Later in the play, as the pilot starts to fall apart mentally, she says that "if I was in a war, a real war, we wouldn't be having this conversation, I wouldn't be fucking exhausted, my head full of grey sitting on a couch talking to an Air Force–approved shrink with my husband, no, I would be having a beer with my boys, I would be shooting pool, I could be cranking music" (56).

47. Megan McCloskey, "The War Room: Daily Transition between Battle, Home Takes Toll on Drone Operators," *Stars and Stripes*, October 27, 2009.

48. Martin and Sasser, *Predator*.

49. The reference is to Creech Air Force base in Nevada; Colin Freeman, "Armchair Killers," *Daily Telegraph,* April 11, 2015.

50. Andrew Niccol, *Good Kill* (IFC Films, 2014).

51. Wood, "Drone Strikes."

52. Greg Jaffe, "Combat Generation: Drone Operators Climb on the Winds of Change in the Air Force," *Washington Post*, February 28, 2010, A1.

53. Ibid.

54. Megan McCloskey, "The War Room: Daily Transition Between Battle, Home Takes a Toll on Drone Operators," *Stars and Stripes*, October 27, 2009.

55. Jane Mayer, "The Predator War," *New Yorker*, October 24.

56. Greg Jaffe, "Combat Generation: Drone Operators Climb on Winds of Change in the Air Force," *Washington Post*, February 28, 2010.

57. Thom Shanker, "A New Medal Honors Drone Pilots and Computer Experts," *New York Times*, February 14, 2013, A16.

58. http://www.duffelblog.com/2012/12/drone-pilot-to-receive-first -air-force-medal-of-honor-since-vietnam/#ixzz3dAfHAqaE.

59. https://www.youtube.com/watch?v=t8-kNPKNCtg.

60. Ernesto Londono, "Pentagon Cancels Divisive Distinguished War-fare Medal for Cyber Ops, Drone Strikes," *Washington Post*, April 15, 2013.

61. Quoted in M. Shane Riza, *Killing without Heart: Limits on Robotic Warfare in an Age of Persistent Conflict* (Dulles, VA: Potomac Books, 2013), 38.

62. My thanks to Kevin Hoder, a former submarine officer (and my brother-in-law) for this point.

63. Singer, *Wired for War*, 331.

64. Chamayou, *A Theory of the Drone*, 101.

65. Glenn Greenwald, "Bravery and Drone Pilots," *Salon,* July 10, 2012.

66. Jonathan Schell, "Attacking Libya—and the Dictionary," *Reader Supported News,* June 21, 2011.

67. Steve Featherstone, "The Coming Robot War," *Harpers*, February 2007, 52.

68. Leif Kaldor and Leslea Mair, *Remote Control War* (Zoot Pictures, 2011).

69. Mayer, "The Predator War."

70. Riza, *Killing without Heart*, 6.

71. Quoted in Mayer, "The Predator War."

72. This passage draws on Hugh Gusterson, "An American Suicide Bomber?," *Bulletin Online*, January 20, 2010; and "Toward an Anthropology of Drones," in Matthew Evangelista and Henry Shue, eds., *The American Way of Bombing: Changing Ethical and Legal Norms*, 191–206 (Ithaca, NY: Cornell University Press, 2014).

73. Talal Asad, *On Suicide Bombing* (New York: Columbia University Press, 2007), 35.

74. Quoted in Kaldor and Mair, *Remote Control War*.

75. Mayer, "The Predator War."

Chapter 3: Remote Intimacy

1. Ron Childress, *And West Is West* (Chapel Hill, NC: Algonquin Books, 2015), 1.

2. Quoted in Sherry Turkle, *Life on Screen: Identity in the Age of the Internet* (New York: Simon and Schuster, 1995), 265.

3. My thanks to Kim Scheppele for this Freudian phrase. Sigmund Freud coined the term *screen memory* to denote false or exaggerated memories from childhood that mask genuine traumatic memories while providing clues to their existence.

4. Ethan Hawke, quoted in Stephanie Merry, "For Hawke, 'Good Kill' Is Good Medicine," *Washington Post*, May 23, 2015, C1.

5. Turkle, *Life on Screen*, 23. See also Turkle's *Alone Together: Why We Expect More from Technology and Less from Each Other* (New York: Basic Books, 2011).

6. Nicola Abé, "Dreams in Infrared," *Der Spiegel Online International*, December 14, 2012.

7. George Brant, *Grounded* (London: Oberon Books, 2013), 43.

8. Craig Whitlock, "When Drones Fall from the Sky," *Washington Post*, June 20, 2014, A1.

9. Peter Singer, *Wired for War: The Robotics Revolution and Conflict in the Twenty-first Century* (New York: Penguin, 2009), 329.

10. Matt J. Martin and Charles W. Sasser, *Predator: The Remote-Control Air War over Iraq and Afghanistan* (Minneapolis, MN: Zenith Press, 2010).

11. Abé, "Dreams in Infrared."

12. In a personal communication, Allison Macfarlane suggests that drone operators' perception of seeing from the plane may be connected to the fact that many of them originally were pilots of manned aircraft and internalized the reflexes of perception in that context. She suggests that drone operators who have not trained as pilots might experience drone imagery differently.

13. Martin and Sasser, *Predator*.

14. Ibid.

15. Liz Wells, *Photography: A Critical Introduction* (New York: Routledge, 2009), 179.

16. Abé, "Dreams in Infrared."

17. Martin and Sasser, *Predator*, 247–248.

18. Colin Freeman, "Armchair Killers: Life as a Drone Pilot," *Daily Telegraph*, April 11, 2015. The British Ministry of Defense made publicly available footage of a British drone strike on an insurgent planting an IED in Afghanistan. It can be seen at https://www.youtube.com/watch?v=GShSMMLooJg.

19. Singer, *Wired for War*, 320.

20. Quoted in Chris Woods, *Sudden Justice: America's Secret Drone Wars* (New York: Oxford University Press, 2015), 86–87.

21. David Wood, "Drone Strikes: A Candid, Chilling Conversation with Top U.S. Drone Pilot," *Huffington Post*, May 15, 2013.

22. Elijah Solomon Hurwitz, "Drone Pilots: 'Overworked, Underpaid, and Bored,'" *Mother Jones*, June 18, 2013.

23. Martin and Sasser, *Predator*, 75–76.

24. Brant, *Grounded*, 41. Brant's seventy-one-page play has no periods.

25. Martin and Sasser, *Predator*, 96.

26. Andrew Cockburn, *Kill Chain: The Rise of the High-Tech Assassins* (New York: Henry Holt, 2015), 200.

27. David S. Cloud, "Anatomy of an Afghan War Tragedy," *Los Angeles Times*, April 10, 2011. This incident also is analyzed in detail in chapter 1 of Cockburn, *Kill Chain*.

28. Ibid.

29. Ibid.

30. Ibid.

31. Ibid.

32. Ibid.

33. Quoted in Cockburn, *Kill Chain*, 16.

34. Ben Emmerson, *Report of the Special Rapporteur on the Promotion and Protection of Human Rights and Fundamental Freedoms While Countering Terrorism* (New York: United Nations Office of the High Commissioner on Human Rights, 2014), 8. A similar hunger to kill is apparent in a video that was leaked by Chelsea Manning. It shows an Apache helicopter crew as it killed civilians that it assumed to be insurgents and can be found at https://www.youtube.com/watch?v=5rXPrfnU3G0.

35. Cockburn, *Kill Chain*, 13.

36. Fellowship of Reconciliation, *Convenient Killing: Armed Drones and the "PlayStation" Mentality*, 2010, https://dronewarsuk.files.wordpress.com/2010/10/conv-killing-final.pdf.

37. Philip Alston and Hina Shamsi, "A Killer above the Law?," *The Guardian*, February 8, 2010.

38. Pratap Chatterjee, "Is Drone Warfare Fraying at the Edges?," *TomDispatch*, March 5, 2015.

39. Jeffrey A. Sluka, "Virtual War in the Tribal Zone: Air Strikes, Civilian Casualties, and Losing Hearts and Minds in Afghanistan and Pakistan," in Neil Whitehead, ed., *Virtual War and Magical Death: Technologies and Imaginaries for Terror and Killing*, 197–193 (Durham, NC: Duke University Press, 2013), 178.

40. David Munoz-Rojas and Jean-Jacques Frésard, *The Roots of Behavior: Understanding and Preventing IHL Violations* (Geneva: International Committee of the Red Cross, 2004), quoted in Klem Ryan, "What's Wrong with Drones? The Battlefield in International Law," in Matthew Evangelista and Henry Shue, eds., *The American Way of Bombing: Changing Ethical and Legal Norms, from Flying Fortresses to Drones*, 207–234 (Ithaca, NY: Cornell University Press, 2014), 213.

41. Fellowship of Reconciliation, *Convenient Killing*, 4.

42. Derek Gregory, "From a View to a Kill: Drones and Late Modern War," *Theory, Culture and Society* 28, no. 7–8 (2011): 188–215.

43. Dave Grossman, *On Killing: The Psychological Cost of Learning to Kill in War and Society* (Boston: Little, Brown, 1995), 118. Grossman's *Killology* website can be found at http://www.killology.com/books.htm.

44. Abé, "Dreams in Infrared."

45. Elaine Scarry, *The Body in Pain: The Making and Unmaking of the World* (New York: Oxford University Press, 1985), 3–4.

46. Interview posted at https://www.youtube.com/watch?v=u2jepIJXHwM, accessed June 10, 2015.

47. Quoted in Chatterjee, "Is Drone Warfare Fraying."

48. Quoted in Gregory, "From a View to a Kill," 198.

49. Antonius C.G.M. Robben, "The Hostile Gaze," in Neil Whitehead, ed., *Virtual War and Magical Death: Technologies and Imaginaries for Terror and Killing* (Durham, NC: Duke University Press, 2013), 132–151.

50. Martin and Sasser, *Predator*.

51. Ibid.

52. Susan Sontag, *Regarding the Pain of Others* (New York: Farrar, Straus and Giroux, 2002), 8.

53. William Broyles Jr., "Why Men Love War," *Esquire*, November 1984.

54. Martin and Sasser, *Predator,* 54

55. Wood, "Drone Strikes."

56. Rachel Martin, "High Levels of 'Burnout' in U.S. Drone Pilots," *All Things Considered*, NPR, December 18, 2011; Elisabeth Bumiller, "Air Force Drone Operators Report High Levels of Stress," *New York Times*, December 18, 2011, A8.

57. Cockburn, *Kill Chain*, 174.

58. Matthew Power, "Confessions of a Drone Warrior," *GQ*, October 23, 2013.

59. Woods, *Sudden Justice*, xvi.

60. Erin Finley, *Fields of Combat: Understanding PTSD among Veterans of Iraq and Afghanistan* (Ithaca, NY: ILR Press, 2011). For the original argument presenting PTSD as a "moral injury," see Jonathan Shay, *Achilles in Vietnam: Combat Trauma and the Undoing of Character* (New York: Athenaeum, 1996).

Chapter 4: Casualties

1. Quoted in Spencer Ackerman, "US Drone Strikes More Deadly to Civilians Than Manned Aircraft," *Then Guardian*, July 2, 2013.

2. Anand Gopal, "Taliban Prisoner Swap Makes Sense," *CNN.com*, June 5, 2014.

3. Details of the conference and links to videos of the presentations can be found at http://conferences.asucollegeoflaw.com/drones/schedule. See http://conferences.asucollegeoflaw.com/drones/videos/panel-iv-drones -and-the-future-of-law for a video of Malinowski's presentation. Malinowski subsequently left Human Rights Watch to become Assistant Secretary of State for democracy, human rights, and labor.

4. Peter Singer, *Wired for War: The Robotics Revolution and Conflict in the Twenty-first Century* (New York: Penguin, 2009), 397–398.

5. Quoted in Scott Shane, "US Said to Target Rescuers at Drone Strike Sites," *New York Times*, February 5, 2012, A4.

6. Glenn Greenwald, "On Media Outlets That Continue to Describe Unknown Drone Victims as 'Militants,'" *The Intercept*, November 18, 2014.

7. Quoted in Jonathan S. Landay, "Obama's Drone War Kills 'Others,' Not Just al Qaida Leaders," *McClatchy News*, April 9, 2013.

8. Dexter Filkins, "What We Don't Know about Drones," *The New Yorker*, February 7, 2013.

9. For more on just war theory, see Oliver O'Donovan, *The Just War Revisited* (New York: Cambridge University Press, 2003); Paul Ramsey, *The Just War* (New York: Scribner's, 1969); Michael Walzer, *Just and Unjust Wars: A Moral Argument with Historical Illustrations* (New York: Basic Books, 1977); Michael Walzer, *Arguing about War* (New Haven, CT: Yale University Press, 2004).

10. James Cavallaro, Stephan Sonnenberg, and Sarah Knuckey, *Living under Drones: Death, Injury, and Trauma to Civilians from US Drone Practices in Pakistan* (Stanford, CA: International Human Rights and Conflict Resolution Clinic, Stanford Law School, and Global Justice Clinic, New York University School of Law, 2012), 4–5, 26–27.

11. Ibid., 33.

12. The methodological difficulties in casualty estimation are discussed at http://www.thebureauinvestigates.com/2011/08/10/pakistan-drone -strikes-the-methodology2; in Cavallaro, Sonnenberg, and Knuckey, *Living under Drones*; and in Alice Ross, "Documenting Civilian Casualties," in Marjorie Cohn, ed., *Drones and Targeted Killing: Legal, Moral and Geopolitical Issues* (Northampton, MA: Olive Branch Press, 2015), 99–131.

13. Greenwald, "On Media Outlets That Continue."

14. Steven Coll, "The Unblinking Stare," *The New Yorker*, November 24, 2014.

15. Cavallaro, Sonnenberg, and Knuckey, *Living under Drones*, 39.

16. http://www.thebureauinvestigates.com/category/projects/drones/drones-graphs. The Bureau of Investigative Journalism also documents casualties from drone strikes in Somalia and Yemen. The Bureau has worked closely with Noor Behram, a Pakistani photojournalist who travels to the sites of drone strikes at considerable personal risk to document their aftermath. Behram's work has been profiled in both *The Guardian* and *The New Yorker*. To give some sense of the difficulty of his task, he says that "there are just pieces of flesh lying around after a strike. You can't find bodies. So the locals pick up the flesh and curse America. ... The youth in the area surrounding a strike gets crazed. Hatred builds up inside those who have seen a drone attack. The Americans think it is working, but the damage they are doing is far greater." Peter Beaumont, "US Drone Strikes in Pakistan Claiming Many Civilian Victims, Says Campaigner," *The Guardian*, July 17, 2011.

17. http://securitydata.newamerica.net/drones/pakistan/analysis.html. For a critique of New America Foundation's overreliance on government sources and failure to update its database as new evidence becomes available, see Cavallaro, Sonnenberg, and Knuckey, *Living under Drones*, 47–52.

18. Ibid., 46–47.

19. http://www.longwarjournal.org/pakistan-strikes. The quote is from Scott Shane, "C.I.A. Is Disputed on Civilian Toll in Drone Strikes," *New York Times*, August 11, 2011, A1.

20. The report can be found at http://www.livingunderdrones.org/download-report. See v and 32. See also David Zucchino, "Drone Strikes in Pakistan Have Killed Many Civilians, Study Says," *Los Angeles Times*, September 24, 2012; Glenn Greenwald, "New Stanford/NYU Study Documents the Civilian Terror from Obama's Drones," *The Guardian*, September 25, 2012.

21. Emmerson's report can be downloaded at http://dronecenter .bard.edu/tag/ben-emmerson. See also Chris Woods, "Leaked Report: High Civilian Death Toll from Drone Strikes," *Salon*, July 22, 2013. Some accounts give slightly lower numbers of children killed in this strike.

22. Spencer Ackerman, "US Drone Strikes More Deadly to Afghan Civilians Than Manned Aircraft—Adviser," *The Guardian*, July 2, 2013; Chris Woods, *Sudden Justice: America's Secret Drone Wars* (New York: Oxford University Press,), 7.

23. Peter Baker, "Obama Apologizes After Drone Kills American and Italian Held by Al Qaeda," *New York Times*, April 23, 2015, A1.

24. Brian Rappert, "The Distribution and Resolution of the Ambiguities of Technology, or Why Bobby Can't Spray," *Social Studies of Science* 31, no. 4 (2001): 557–591. See also Cheryl W. Thompson and Mark Berman, "Improper Techniques, Increased Risks," *Washington Post*, November 26, 2015, 1.

25. Spencer Ackerman, "Drone Strikes by UK and Pakistan Point to Obama's Counter-terror Legacy," *Guardian*, September 9, 2015.

26. Charles Krauthammer, "Barack Obama: Drone Warrior," *Washington Post*, May 31, 2012; Jeremy Scahill, "The Assassination Complex," *The Intercept*, October 15, 2015. For a rough approximation of a "baseball card," see Josh Begley, "A Visual Glossary," *The Intercept*, October 15, 2015.

27. See Cavallaro, Sonnenberg, and Knuckey, *Living under Drones*, 14–15.

28. Scahill, "Assassination Complex;" Cora Currier, "The Kill Chain," *The Intercept*, October 15, 2015.

29. Adam Entous, "Special Report: How the White House Learned to Love the Drone," *Reuters*, May 18, 2010. A similar conclusion is reached by Peter Bergen and Megan Braun, "Drone Is Obama's Weapon of Choice," *CNN*, September 6, 2012.

30. Kevin Jon Heller, "'One Hell of a Killing Machine': Signature Strikes and International Law," *Journal of International Criminal Justice* 11, no. 1 (2013): 89–119.

31. Coll, "The Unblinking Stare."

32. Daniel Klaidman, *Kill or Capture: The War on Terror and the Soul of the American Presidency* (New York: Houghton Mifflin Harcourt, 2012).

33. Andrew Cockburn, *Kill Chain: The Rise of the High-Tech Assassins* (New York: Henry Holt, 2015), 224.

34. Klem Ryan, "What's Wrong with Drones? The Battlefield in International Law," in Matthew Evangelista and Henry Shue, eds., *The American Way of Bombing: Changing Ethical and Legal Norms, from Flying Fortresses to Drones* (Ithaca, NY: Cornell University Press, 2014), 207–223.

35. David S. Cloud, "Anatomy of an Afghan War Tragedy," *Los Angeles Times*, April 10, 2011.

36. Jo Becker and Scott Shane, "Secret 'Kill List' Proves a Test of Obama's Principles and Will,'" *New York Times*, May 29, 2012, A1.

37. Landay, "Obama's Drone War Kills 'Others.'"

38. Becker and Shane, "Secret 'Kill List.'"

39. Micah Zenko, "If Trayvon Were Pakistani ...," *Foreign Policy*, July 22, 2013. For a similar argument, see Mark LeVine, "Obama's Haunting and the Curse of Governance," *Al Jazeera*, May 28, 2013. Trayvon Martin was a 17-year-old African American who was shot dead in 2012 while walking through a gated community in Florida, where he was staying with a relative. The community had experienced a spate of robberies and Martin looked suspicious to a neighborhood-watch volunteer who got into an altercation with him and shot him, saying he felt threatened by him. The case excited considerable media coverage, and Martin was widely portrayed as a victim of unjust racial profiling. For President Obama's remarks on the incident, which foregrounded the issue of racial profiling, see https://www.whitehouse.gov/the-press-office/2013/07/19/remarks-president-trayvon-martin.

40. See Conor Friedersdorf, "Drone Attacks at Funerals of People Killed in Drone Attacks," *The Atlantic*, October 24, 2013. Friedersdorf quotes a drone operator who writes in a Facebook discussion of posttraumatic stress disorder among drone operators, "How many of you have killed a group of people, watched as their bodies are picked up, watched the funeral, then killed them too?"

41. Scott Shane, "Report Cites High Civilian Toll in Pakistan Drone Strikes," *New York Times*, September 25, 2012; Ryan, "What's Wrong with Drones?," 219; Cavallaro, Sonnenberg, and Knuckey, *Living under Drones*, 75.

42. Glenn Greenwald, "US Drone Strikes Target Rescuers in Pakistan— and the West Stays Silent," *The Guardian*, August 20, 2012; Glenn Greenwald, "US Drones Targeting Rescuers and Mourners," *Salon*, February 5, 2012.

43. Cavallaro, Sonnenberg, and Knuckey, *Living under Drones*, 76.

44. "Obama's Lists: A Dubious History of Targeted Killings in Afghanistan," *Der Spiegel*, December 28, 2014.

45. Reprieve, "You Never Die Twice: Multiple Kills in the US Drone Program," November 2014, http://www.reprieve.org/uploads/2/6/3/3/26338131/2014_11_24_pub_you_never_die_twice_-_multiple_kills_in_the_us_drone_program.pdf.

46. Landay, "Obama's Drone War Kills 'Others.'"

47. Woods, *Sudden Justice*, 56.

48. "Obama's Lists."

49. Jeremy Scahill and Glenn Greenwald, "The NSA's Secret Role in the U.S. Assassination Program," *The Intercept*, February 10, 2014, https://firstlook.org/theintercept/2014/02/10/the-nsas-secret-role.

50. Scahill, "The Assassination Complex."

51. Cockburn, *Kill Chain*, 197. See also Quil Lawrence, "Afghan Raids Common, But What If Target Is Wrong?," *NPR Morning Edition*, May 12,

2011; Julius Cavendish, "Intelligence Failures 'Led to Deaths of Afghan Civilians,'" *The Independent*, May 12, 2011.

52. "Obama's Lists."

53. AP, "Ex-Bush Official: Many at Guantanamo Bay Are Innocent," *FoxNews.com*, March 19, 2009, http://www.foxnews.com/politics/2009/03/19/ex-bush-official-guantanamo-bay-innocent.

54. Akbar Ahmed, *The Thistle and the Drone: How America's War on Terror Became a Global War on Tribal Islam* (Washington, DC: Brookings Institution Press, 2013), 82–83.

55. Cockburn, *Kill Chain*, 228.

56. Anand Gopal, *No Good Men among the Living: America, the Taliban, and the War through Afghan Eyes* (New York: Metropolitan Books, 2014), 109–110.

57. Conor Friedersdorf, "Official: Team Bush Knew Many at Gitmo Were Innocent," *The Atlantic*, April 26, 2013; and https://www.aclu.org/infographic/guantanamo-numbers.

58. Susanne Koelbl, "NATO High Commander Issues Illegitimate Order to Kill," *Der Spiegel*, January 28, 2009; Matthias Gebauer and Susanne Koelbl, "Order to Kill Angers German Politicians," *Der Spiegel*, January 29, 2009.

59. Jonathan S. Landay, "U.S. Secret: CIA Collaborated with Pakistan Spy Agency in Drone War," *McClatchy News*, April 9, 2013.

60. Mark Mazzetti, "A Secret Deal on Drones, Sealed in Blood," *New York Times*, April 6, 2013, A1. See also Mark Mazzetti, *The Way of the Knife: The CIA, a Secret Army, and a War at the Ends of the Earth* (New York: Penguin Books, 2013).

61. Many observers have claimed that operational protocols for drone strikes differ depending on whether the drones are operated by the U.S. military or the CIA. Drone strikes over Afghanistan are undertaken by the U.S. military, and drone strikes over Pakistan tend to be a CIA initiative, though JSOC is also involved. Many have claimed that the CIA is

more secretive about its targeting procedures. In the talk by Tom Malinowski of Human Rights Watch that opens this chapter, Malinowski argued that it would be best if responsibility for all drone strikes was consolidated under the U.S. military.

62. Mazzetti, "A Secret Deal on Drones."

63. Ibid.

64. Jeremy Scahill, *Dirty Wars: The World Is a Battlefield* (New York: Nation Books, 2014).

65. Ryan Devereaux, "Manhunting in the Hindu Kush," *The Intercept*, October 15, 2015.

66. John Brennan, "The Efficacy and Ethics of U.S. Counterterrorism Strategy," Woodrow Wilson Center, Washington DC, April 30, 2012.

67. See Ahmed, *The Thistle and the Drone*; *Meeting Resistance*, directed by Steve Connors and Molly Bingham, Nine Lives Documentary Productions, 2007, film; Gopal, *No Good Men among the Living*; Patrick Graham, "Beyond Fallujah: A Year with the Iraqi Resistance," *Harpers*, June 2004, 37–48; Dahr Jamail, *Beyond the Green Zone: Dispatches from an Unembedded Journalist in Occupied Iraq* (Chicago: Haymarket Books, 2007); David Kilcullen, *The Accidental Guerilla: Fighting Small Wars in the Midst of a Big One* (New York: Oxford University Press, 2011); Tom Roberts, *The Insurgency*, PBS Frontline, 2006; Nir Rosen, "How We Lost the War We Won," *Rolling Stone*, October 30, 2008; Anthony Shadid, *Night Draws Near: Iraq's People in the Shadow of America's War* (New York: Picador, 2006).

68. Coll, "The Unblinking Stare."

69. Stanley McCrystal interview on *Today* programme, BBC radio January 24, 2001, quoted in Woods, *Sudden Justice*, 9.

70. Kilcullen, *The Accidental Guerilla.*

71. Becker and Shane, "Secret 'Kill List.'"

72. Gregory Johnson, interview on *All Things Considered*, NPR September 11, 2014 (transcribed by author); Gregory Johnsen, *The Last Refuge: Yemen, el Qaeda, and America's War in Arabia* (New York: Norton, 2012).

73. Ibrahim Mothana, "How Drones Help al Qaeda," *New York Times*, June 13, 2012, A35.

74. Klaidman, *Kill or Capture*, 119.

75. Cockburn, *Kill Chain*, 167.

76. Rafiq ur Rehman, "Please Tell Me, Mr. President, Why a U.S. Drone Assassinated My Mother," *Guardian*, October 25, 2013.

77. Coll, "The Unblinking Stare."

78. Ibid.

79. Ibid.

80. https://www.whitehouse.gov/the-press-office/2013/05/23/remarks-president-national-defense-university.

81. Steven Coll, "Warren Weinstein and the Long Drone War," *The New Yorker*, April 23, 2015.

Chapter 5: Arsenal of Democracy?

1. Adam Kirsch, "Review of *The Illuminations* by Andrew O'Hagan," *Washington Post*, April 8, 2015.

2. Scott Shane, "Drone Strikes Reveal Uncomfortable Truth: U.S. Is Often Unsure about Who Will Die," *New York Times*, April 24, 2015, A1. On the ACLU requests, see Spencer Ackerman, "ACLU Files Lawsuit over Obama Administration 'Kill List,'" *The Guardian*, March 16, 2015.

3. Harold Koh, speech at the Oxford Union, May 7, 2013, as reported in Conor Friedersdorf, "Harold Koh's Slippery, Inadequate Criticism of the Drone War," *The Atlantic*, May 9, 2013.

4. Remarks by President Barack Obama at the National Defense University, May 23, 2013, https://www.whitehouse.gov/the-press-office/2013/05/23/remarks-president-national-defense-university.

5. Eric Holder speech at Northwestern University Law School, March 5, 2012, http://www.justice.gov/opa/speech/attorney-general-eric-holder-speaks-northwestern-university-school-law.

6. John Brennan, "The Efficacy and Ethics of U.S. Counterterrorism Strategy," Woodrow Wilson Center, Washington, DC, April 30, 2012.

7. Harold Koh, "The Obama Administration and International Law," speech at the Annual Meeting of the American Society of International Law, Washington, DC, March 25, 2010, http://www.state.gov/s/l/releases/remarks/139119.htm.

8. See https://rethinkkoh.wordpress.com for the full text of the petition. My thanks to Richard Wilson for bringing this to my attention.

9. Koh made this comment after he left office in his May 7, 2013, speech at the Oxford Union (see note 3). *Jus ad bellum* refers to the ethics of declaring a particular war. *Jus in bello* refers to the ethics of actions undertaken within a war.

10. Brennan responded thus to the reporter, implicitly conceding that his account of procedures for selecting targets did not apply to all cases: "You make reference to signature strikes that are frequently reported in the press. I was speaking here specifically about targeted strikes against individuals."

11. The argument that the Obama administration has preferred to kill rather than capture insurgent leaders has been made forcefully in Mark Mazzetti, *The Way of the Knife: The CIA, a Secret Army, and a War at the Ends of the Earth* (New York: Penguin, 2013); and Daniel Klaidman, *Kill or Capture: The War on Terror and the Soul of the Obama Presidency* (New York: Houghton Mifflin, Harcourt, 2012).

12. Al-Awlaki's whereabouts were well known in Yemen. According to an article in *The Guardian*, no request was made to arrest him. See

Ghaith Abdul-Ahad, "Shabwa: Blood Feuds and Hospitality in al-Qaida's Yemen Outpost," *The Guardian*, August 23, 2010.

13. Evan Perez, "Judge Dismisses Targeted-Killing Suit," *Wall Street Journal*, December 8, 2010.

14. John Kaag and Sarah Kreps, *Drone Warfare* (Cambridge, UK: Polity, 2014), 60.

15. Ibid., 61–63; Jaime Fuller, "Americans Are Fine with Drone Strikes: The Rest of the World? Not So Much," *Washington Post*, July 15, 2014; Bruce Drake, "Report Questions Drone Use, Widely Unpopular Globally, But Not in the U.S.," *Pew Research Center*, October 23, 2013. One poll suggests that U.S. support for drone strikes drops to 29 percent if there are civilian casualties. See Emily Swanson, "Drone Poll Finds Support for Drone Strikes, with Limits," *Huffington Post*, August 2, 2013.

16. Philip Alston, special rapporteur on extrajudicial, summary, or arbitrary executions, *Study on Targeted Killings*, 2010, http://www2.ohchr .org/english/bodies/hrcouncil/docs/14session/A.HRC.14.24.Add6.pdf; Ben Emmerson, *Report of the Special Rapporteur on the Promotion and Protection of Human Rights and Fundamental Freedoms While Countering Terrorism*, 2014, https://www.justsecurity.org/wp-content/uploads/ 2014/02/Special-Rapporteur-Rapporteur-Emmerson-Drones-2014.pdf; Christof Heyns, *Report of the Special Rapporteur on Extrajudicial, Summary or Arbitrary Executions*, 2013, https://www.justsecurity.org.

17. James Cavallaro, Stephan Sonnenberg, and Sarah Knuckey, *Living under Drones: Death, Injury, and Trauma to Civilians from U.S. Drone Practices in Pakistan* (Stanford, CA: International Human Rights and Conflict Resolution Clinic, Stanford Law School, and Global Justice Clinic, New York University School of Law, 2012).

18. Amnesty International, *Will I Be Next? US Drone Strikes in Pakistan*, 2013, http://www.amnestyusa.org/research/reports/will-i-be-next-us -drone-strikes-in-pakistan.

19. See https://www.thebureauinvestigates.com/category/projects/drones and http://www.reprieve.org.uk/topic/drones.

20. See Medea Benjamin, *Drone Warfare: Killing by Remote Control* (New York: Verso, 2013); Andrew Cockburn, *Kill Chain: The Rise of the High-Tech Assassins* (New York: Henry Holt); Nick Turse, *Terminator Planet: The First History of Drone Warfare, 2001–2050* (Lexington, KY: Dispatch Books, 2012); Chris Woods, *Sudden Justice: America's Secret Drone Wars* (New York: Oxford University Press, 2015).

21. Salman Masood and Ihsanullah Tipu Mehsud, "Thousands in Pakistan Protest American Drone Strikes," *New York Times*, November 24, 2013, A10.

22. A dramatic video of Benjamin's interruption of Brennan's speech can be found at https://www.youtube.com/watch?v=H6hBjDlPIbQ.

23. Quoted in Gregoire Chamayou, *A Theory of the Drone* (New York: New Press, 2013), 167.

24. Alston, *Study on Targeted Killings*, 3.

25. James Joyner, "Killing Americans," *National Interest*, February 6, 2013.

26. Heyns, *Report*, 23.

27. Charlie Savage and Mark Landler, "War Powers Act Doesn't Apply for Libya, Obama Says," *New York Times*, June 15, 2011, A16.

28. Jack Goldsmith, "Obama's Breathtaking Expansion of a President's Power to Make War," *Time*, September 11, 2014; Savage and Landler, "War Powers Act."

29. Authorization for Use of Military Force against Terrorists, 115 Stat. 224 (2001).

30. Quoted in Kaag and Kreps, *Drone Warfare*, 66.

31. Scott Barber, "Before He Was Overthrown and Killed, Libyan Dictator Muammar Gadaffi Warned Jihadists Would Conquer Northern Africa," *National Post*, January 25, 2013.

32. Both quotes are from Eli Lake, "Obama's New War on ISIS May Be Illegal, *Daily Beast*, September 14, 2014.

33. David Dow, "In Assassinating Anwar al-Awlaki, Obama Left the Constitution Behind," *Daily Beast*, March 16, 2012.

34. Ibid.

35. David Cole, "The Drone Memo: Secrecy Made It Worse," *New York Review of Books blog*, June 24, 2014.

36. Dow, "In Assassinating Anwar al-Awlaki."

37. Ibid.

38. David Cole, "President Obama, Did You or Did You Not Kill Anwar al-Awlaki?," *Washington Post*, February 8, 2013.

39. Glenn Greenwald, "Excuses for Assassination Secrecy," *Salon*, July 12, 2012.

40. Brian Glynn Williams, *Predators: The CIA's Drone War on al Qaeda* (Lincoln, NB: Potomac Books, 2013), 115. Cavallaro, Sonnenberg, and Knuckey, *Living under Drones*, 15, cite a U.S. embassy cable made public by Wikileaks in which Pakistan's prime minister is quoted as saying that he has no problem with drone strikes "as long as they get the right people. We'll protest in the National Assembly and then ignore it."

41. Williams, *Predators*, 115.

42. Emmerson, *Report of the Special Rapporteur*, 15.

43. Heyns, *Report*, 19. The quote is from the 1842 Caroline case.

44. NBC News, "Justice Department Memo Reveals Legal Case for Drone Strikes on Americans," February 4, 2013; Jeffrey Rosen, "Drone Strike Out," *New Republic*, February 6, 2013.

45. *Last Night with John Oliver*, September 28, 2014, HBO, https://www .youtube.com/watch?v=K4NRJoCNHIs.

46. Emmerson, *Report of the Special Rapporteur*, 18, 20.

47. Ibid., 19.

48. Alston, *Study on Targeted Killings*, 18.

49. Woods, *Sudden Justice*, 93.

50. Ibid., 18.

51. Heyns, *Report*, 16, 23.

52. Ibid., 15.

53. Alston, *Study on Targeted Killings*, 24.

54. Heyns, *Report*, 5.

55. Kaag and Kreps, *Drone Warfare*, 108.

56. Immanuel Kant, "Towards Perpetual Peace," *The Basic Writings of Kant*, edited by Allen Wood (New York: Random House, 2001), 422.

57. Quoted in M. Shane Riza, *Killing without Heart: Limits on Robotic Warfare in an Age of Persistent Conflict* (Washington, DC: Potomac Books, 2013), 77.

58. Chamayou. *A Theory of the Drone*, 188.

59. Kaag and Kreps, *Drone Warfare*, 76–77.

60. Stephen Holmes, "What's in It for Obama?," *London Review of Books*, July 18, 2013.

61. Chamayou, *A Theory of the Drone*, 111.

62. Mary Kaldor, *New and Old Wars: Organized Violence in a Global Era* (Stanford, CA: Stanford University Press, 2007). In the second edition of the book, Kaldor concedes to her critics the difficulty of knowing who is and is not a civilian and therefore of precisely estimating the ratio of civilian casualties. I believe that her general point still stands that the twentieth century saw a shift in war that threatened to erase the combatant-civilian distinction.

63. Talal Asad, *On Suicide Bombing* (New York: Columbia University Press, 2007), 35.

64. Carl von Clausewitz, *On War* (London: Penguin, 1982 [1832]), 101.

65. Elaine Scarry, *The Body in Pain: The Making and Unmaking of the World* (New York: Oxford University Press, 1985), 89.

66. Chamayou, *A Theory of the Drone*, 13.

67. Ibid., 162–163.

68. Ibid., 33–34.

69. Ibid., 32.

70. Ibid., 62–63.

71. Ibid., 53.

Conclusion

1. Quoted in Scott Shane, "Coming Soon: The Drone Arms Race," *New York Times*, October 9, 2011, SR5.

2. According to the U.S. Government Accountability Office, "at least 76 countries" already have some sort of drone capability. Iran has developed a drone called "the Ambassador of Death." The Pentagon has given permission for drone exports to sixty-six countries. James Cavallaro, Stephan Sonnenberg, and Sarah Knuckey, *Living under Drones: Death, Injury, and Trauma to Civilians from U.S. Drone Practices in Pakistan* (Stanford, CA: International Human Rights and Conflict Resolution Clinic, Stanford Law School, and Global Justice Clinic, New York University School of Law, 2012), 141–142.

3. Salman Masood, "Pakistan Says Its Drone Killed Three Militants," *New York Times*, September 8, 2015, A10.

4. David Hambling, "Air Force Completes Killer Micro-Drone Project," *Wired*, January 5, 2010; Mike Adams, "Citizens Strike Back: Tiny, Low-

Cost Drones May One Day Assassinate Corrupt Politicians, Corporate CEOs and Street Criminals," *Natural News*, June 9, 2014, http://www .naturalnews.com/045495_assassination_drones_autonomous_killing _facial_recognition.html; Peter Finn, "A Future for Drones: Automated Killing," *Washington Post*, September 19, 2011.

5. Craig Whitlock, "Near Collisions between Drones, Airliners Surge, New FAA Reports Show," *Washington Post*, November 26, 2014.

6. Nicholas Rondinone and Kathleen Megan, "CCSU Professor to Eighteen-Year-Old: Making Gun-Firing Drone a 'Terrible Idea,'" *Hartford Courant*, July 22, 2015, http://www.courant.com/breaking-news/hc-gun -fire-drone-investigation-20150721-story.html.

7. http://icrac.net/call.

8. Steven Nelson, "Weaponized, Peeping Drone Ban Proposed in Congress," *US News and World Report*, March 17, 2015.

9. Ken Butigan, "Envisioning an International Treaty Banning Drones," *Waging Nonviolence*, May 23, 2013, http://wagingnonviolence .org/feature/envisioning-an-international-treaty-banning-drones/; David Swanson, "Fifty Organizations Seek Ban on Armed Drones," *Roots Action*, November 10, 2013. See also http://www.commondreams.org/ newswire/2014/10/01/ban-weaponized-drones-international-day -action-october-4th.

10. RT News, "No to Killer Drones; UN Chief Calls for UAV Surveillance Use Only," *RT News*, August 13, 2013.

11. Harold Koh, speech at the Oxford Union, May 7, 2013, as reported in Conor Friedersdorf, "Harold Koh's Slippery, Inadequate Criticism of the Drone War," *The Atlantic*, May 9, 2013.12. Mark Mazzetti, *The Way of the Knife: The CIA, a Secret Army, and a War at the Ends of the Earth* (New York: Penguin, 2013); Daniel Klaidman, *Kill or Capture: The War on Terror and the Soul of the Obama Presidency* (New York: Houghton Mifflin Harcourt, 2012).

13. Stephen Holmes, "What's in It for Obama?," *London Review of Books*, July 18, 2013.

14. David Kilcullen, *The Accidental Guerilla: Fighting Small Wars in the Midst of a Big One* (New York: Oxford University Press, 2009); David Kilcullen, *Counterinsurgency* (New York: Oxford University Press, 2010); John Nagl, *Learning to Eat Soup with a Knife: Counterinsurgency Lessons from Malaya and Vietnam* (Chicago: University of Chicago Press).

Index